Construction Cost Management

In the last decade, following the Latham and Egan reports, there have been many significant changes in the role of the construction cost manager. Keith Potts examines the key issues and best practice in the cost management of construction projects under traditional contracts and new methodologies. All stages within the life cycle of a project are considered from pre-contract to tendering and post-contract.

Worked examples, legal cases and over 65 project case studies are used to illustrate the practical application of the theory, where appropriate. Extensive references are captured, including the UK government's Constructing Excellence programme and the National Audit Reports, in order to further develop an understanding of the subject. Reference is made to major projects such as the Scottish Parliament Building, Wembley Stadium and BAA's Heathrow Terminal 5.

Aimed at students of Surveying and Construction Management programmes, this book uniquely embraces cost management in both the building and civil engineering sectors in the UK and overseas and should thus prove useful to practitioners. Seminar questions are included at the end of each chapter with additional links to over 100 project case studies in order to reinforce the learning experience.

Keith Potts is Senior Lecturer in the School of Engineering and the Built Environment at the University of Wolverhampton, is a RICS external examiner in Quantity Surveying, and Award Leader of the RICS-accredited MSc in Construction Project Management.

Also available from Taylor & Francis

Construction Cost Management

Learning from case studies

Keith Potts

Taylor & Francis
Taylor & Francis Group

LONDON AND NEW YORK

First published 2008
by Taylor & Francis
2 Park Square, Milton Park, Abingdon, Oxon OX14 4RN

Simultaneously published in the USA and Canada
by Taylor & Francis
270 Madison Ave, New York, NY 10016

*Taylor & Francis is an imprint of the Taylor & Francis Group,
an informa business*

© 2008 Keith Potts

Typeset in Frutiger Light by
Newgen Imaging Systems (P) Ltd, Chennai, India
Printed and bound in Great Britain by
TJ International, Padstow, Cornwall

British Library Cataloguing in Publication Data
A catalogue record for this book is available from the British Library

Library of Congress Cataloging in Publication Data
Potts, Keith F.
  Construction cost management: learning from case studies /
  Keith Potts.
    p. cm.
  Includes bibliographical references and index.
   1. Construction industry – Costs.
  2. Construction industry – Cost control. I. Title.

TA682.26.P68 2008
624.068′1–dc22   2007032917

ISBN10: 0–415–44286–9 (hbk)
ISBN10: 0–415–44287–7 (pbk)
ISBN10: 0–203–93301–X (ebk)

ISBN13: 978–0–415–44286–2 (hbk)
ISBN13: 978–0–415–44287–9 (pbk)
ISBN13: 978–0–203–93301–5 (ebk)

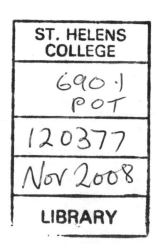

This book is dedicated to the memory of Francis Leon Potts, who first encouraged me to go into quantity surveying, and to the love and friendship of Lesley, Ian, Gemma and Debbie.

Contents

Preface

My first text book *Major Construction Works: Contractual and Financial Management* (Longman) (1995) was based on documentation assembled during 1993 and 1994; this now seems like a prehistoric era before the Latham (1994) and Egan (1998) reports!

In that book, I attempted to identify the key issues in the successful contractual and financial management on major projects. It was based on my experience as a senior quantity surveyor, at separate times employed by both contractor and client, on the Hong Kong Mass Transit Railway – at the time one of the largest construction projects in the world. In contrast to the norm at that time within the UK, the massive Hong Kong project – despite major difficulties – was completed on time and within budget. The lessons to be learned from this project were identified in the case study in the last chapter of the book. It was clear from this experience that any project could be completed on time and within budget providing the appropriate procurement systems, planning and control methods, contracts and financial procedures were in place – crucially with experienced, motivated people to implement them.

Over a decade later, the world seems very different, yet the same fundamentals apply – clients wish to obtain their project within budget and within time and to the necessary quality.

The relentless growth of the World Wide Web (www) meant that all could now easily access a vast array of important information. The problem for students, however, was in identifying which information was significant and which was superfluous.

In this new text, I have attempted to embrace the recommendations of the key reports and government bodies including the National Audit Office and the Office of Government Commerce. The book includes the tools and techniques required under the new partnering/alliancing philosophies as well as including chapters on valuing variations and claims based on the traditional procurement approach. Observations in the book are reinforced throughout with detailed analysis of over 60 project case studies with additional links to over 100 case studies. Many of the project case studies are taken from the *Building* magazine or the *National Audit Office* reports to whom the author is most grateful for permission to publish.

A chapter is included on the NEC *ECC Contract*, which has been the standard contract in the civil engineering and infrastructure sectors for some time and is increasingly chosen by public clients in the building sector. Its choice by the London 2012 Olympic Development Authority reinforces its status. A chapter on the new FIDIC contract is included for those working on major projects outside the UK. Uniquely, the new textbook embraces both the building and civil engineering sectors and should be of interest to both undergraduate and postgraduate students as well as practitioners.

A significant case study on the Heathrow Terminal 5 has been included. It is important that the lessons learned on this pioneering project in lean-construction are disseminated and understood by all.

Over the past two years I have received useful information from many senior quantity surveyors and commercial managers representing both consultants, public and private clients and contractors. These together with comments and observations received from undergraduate and postgraduate students at the University of Wolverhampton have deeply enriched the study. Finally, I wish to thank fellow colleagues Rod Gameson, Chris Williams and particularly Pauline Corbett for their valued help in passing learned comments on the draft chapters. Any errors or omissions are, of course, my responsibility.

Part I
Introduction

1 Introduction and overview

1.1 Setting the scene

There have been many significant changes in the construction sector within the past decade. The relentless development of computer power and the growth of the World Wide Web and knowledge management, the increasing use of Private Finance Initiative (PFI) and Public–Private Partnership (PPP), the growth of partnering and alliancing, the increasing importance of supply-chain management and the increasing use of the *New Engineering Contract* have changed the industry for the good. Yet, the same fundamentals apply – clients wish to obtain their project within budget and within time and to the necessary quality.

Significantly, the role of the quantity surveyor (QS) has also changed and many have moved on from contractual and financial management of projects to embrace the key role as the client's construction manager/project manager. One of the pioneer QS construction project managers was Francis Graves, who undertook the task of project controller in 1972 on the massive five-year-long Birmingham NEC Exhibition Centre project. He considered his terms of reference on this project very straightforward – *Get it finished on time and get value for money!*

An analysis of three of the top QS Consultants' websites shows their involvement in a wide range of cost management and related services (Table 1.1).

Significantly, following their successful partnership on the Heathrow Terminal 5 project, EC Harris and Turner & Townsend established Nuclear JV to tender for programme controls services and to subsequently target the wider nuclear decommissioning programme. Of particular interest, Rob Smith, Chairman of Davis Langdon & Seah, in his introduction to their 2005–2006 *Global Review* identifies some of the big issues facing us all – particularly environmental and sustainability concerns. He comments that part of Davis Langdon's contribution will be the introduction of sustainable design thinking and sustainable metrics into their cost plans (www.davis-langdon.com/mainpage/GlobalReview.htm. [accessed 28 February 2007]).

1.2 Construction overview

The construction sector is strategically important for Europe, providing the infrastructure and buildings on which all sectors of the economy depend. With 11.8 million operatives directly employed in the sector, it is Europe's largest industrial employer accounting for 7% of total employment and 28% of industrial employment in the EU. It is estimated that 26 million workers in the EU depend in one way or another on the construction sector. About €910 billion

Table 1.1 Range of services offered by leading construction cost consultants.

Company	Cost management services
EC Harris	*Consulting*: Asset & Facilities Strategies; Enabling Technology & Management Information Systems; Performance Management; Technical & Investment Risk Management; Public Private Finance Consultancy; Supply-Chain Management; Risk Management & Business Continuity *Managing Delivery*: Programme, Project & Construction Management; Commercial Management, Quantity Surveying & Value Engineering; Safety, Health & Environmental Management; Asset & Facilities Management; Independent Advisor, Due Diligence & Insolvency; Strategic Procurement, Contractual Advice & Dispute Resolution; Development Management *Delivery of Business Support Maintenance & Construction*: Facilities Outsourcing; Prime Contracting
Turner & Townsend	*Inception*: Feasibility Studies; Business Case; Development; Strategic Cost Planning; Benchmarking *Development*: Contract Strategies; Option Appraisals; Capital and Operational Cost Planning; Risk Management; Value Management; Whole-Life Costing; Supplier & Contractor Procurement *Implementation*: Cost Control & Management; Change Control; Payment Mechanisms; Performance Measurement *Operation and Occupation*: Maintenance Strategies; Energy and Cost-in-Use Analysis; Facilities Management
Franklin + Andrews	Benchmarking Employer's Agent Bills of Quantities Feasibility Studies Building Services (M&E) Project Audits Capital Allowances Quantity Surveying/Cost Commercial Management Real Estate Due Diligence Contract Advisory Services Real Estate Tax Management Contract Documentation Whole-Life Costing Analysis Cost Modelling and Planning Due Diligence

Source: Websites www.echarris.com; www.turnerandtownsend.com; www.franklinandrews.com (accessed 28 February 2007).

was invested in construction in 2003, representing 10% of the GDP and 51.2% of the Gross Fixed Capital Formation of the EU (www europa.eu.int [accessed September 2006]).

Moreover, the relationship between construction activities and the built environment, on the one hand, and sustainable development, on the other, is both significant and complex. Construction uses more raw materials than any other sector, and the creation and operation of the built environment accounts for an important consumption of natural resources. There is also a pressing need to address the regeneration of many urban areas of Europe, in particular in the newly acceded countries, and the realization of major trans-European infrastructure works.

The UK construction activity makes a considerable contribution to the national economy accounting for over 8% of the national gross domestic product. UK annual public sector construction output has grown by over a third between 1999 and 2003 from just under £24 billion per year to around £33.5 billion, and capital investment is set to continue expanding over the next three years in key sectors such as schools, hospitals, roads and social housing (NAO, 2005).

Table 1.2 shows the top ten clients/promoters by turnover from 1 January 2006 to 31 December 2006 in the UK based on analysis of contracts with a value of £100,000 or above, and all work awarded by negotiation or competitive tender, including residential and civil work. Significantly, these major clients have shown the lead in embracing the new procurement routes and conditions of contract.

Table 1.2 Top clients/promoters in the UK – year ending 31 December 2006.

Company	Total value (£ million)	Main procurement type
1 British Nuclear Fuels	6,506	Framework agreements Competitive supply chains Fixed price or Target cost contracts (when scope and risk reasonably defined)
2 Partnership for Schools	4,000	Long-term partnerships Private Finance Initiatives Design and build contracts
3 Defence Estates	3,160	Prime contracting Public–Private Partnerships/Private Finance Initiatives
4 Birmingham Council	2,310	500 million 5-year Partnership with 3 construction companies for all projects valued at more than £100,000 Contracts based on NEC3
5 Olympic Delivery Authority	1,565	Based on OGC's *Achieving Excellence in Construction* Guides Projects managed using a delivery partner Contracts based on NEC3

Source: Developed based on *Building* magazine, 26 January 2007, p. 20 and websites: www.britishnucleargroup.com; www.p4s.org.uk/; www.defence-estates.mod.uk/major_projects/ index.htm; www.birmingham.gov.uk; www.london 2012.com (accessed 1 March 2007).

1.3 Influential reports

In the last 60 years there have been numerous reports on the state of the UK construction industry (Murray and Langford, 2003), including the following:

- *Placing and Management of Building Contracts: The Simon Committee Report* (1944);
- *The Working Party Report to the Minister of Works: The Phillips Report on Building* (1948–1950);
- *Survey of Problems before the Construction Industry: A Report Prepared by Sir Harold Emmerson* (1962);
- *The Placing and Management of Contracts for Building and Civil Engineering Work: The Banwell Report* (1964);
- *Tavistock Studies into the Building Industry: Communications in the Building Industry* (1965) and *Interdependence and Uncertainty* (1996);
- *Large Industrial Sites Report* (1970);
- *The Public Client and the Construction Industries: The Wood Report* (1975);
- *Faster Building for Industry: NEDO* (1983);
- *Faster Building for Commerce* (1988).

With the exception of the Tavistock Studies, these were all government-sponsored reports, produced by large committees. Most made recommendations for improvement but were mainly ignored due to poor client involvement and no follow-up legislation to reinforce their findings. Significantly, Banwell (1964) recommended that a common form of contract be used for all construction work. This only started to become a reality 30 years later – with the introduction of the *New Engineering Contract* (NEC).

Thankfully, Sir Michael Latham's *Constructing the Team* (1994) was not ignored, being the catalyst, which has attempted to transform the industry. Since this date we have seen the publication of many significant government-sponsored or National Audit Office Reports including The Egan report *Rethinking Construction* (1998); *Modernising Construction* (2001); The Second Egan report *Accelerating Change* (2002); *Achieving Excellence in Construction Procurement Guides* (2004) and *Improving Public Services through Better Construction* (2005).

These reports and other relevant key recommendations are reviewed in Chapter 2 in order that best practice be identified.

1.4   Recommendations from professional bodies

The RICS APC/ATC requirements and competencies (RICS, 2002)

The RICS identifies that construction surveyors (quantity surveyors) may be working as consultants or working for a contracting or engineering company in the following areas:

- Preparing development appraisals for different sites, assessing the effects of capital and revenue expenditure, life-cycle costs, grants and taxation implications;
- Advising clients on the project brief, preferred procurement routes, costs and cash flow;
- Analysing the whole-life costs of a project;
- Planning the construction process;
- Monitoring and control of cost during the pre-contract stages;
- Preparing tender and contractual documentation, leading to tender selection and appraisal;
- Following the letting of the contract for the project, advising on payments to contractors and post-contract cost control, settlement of final account;
- Controlling a project on behalf of their employer;
- Negotiating with the client or subcontractors;
- Reporting on programme and financial matters;
- Risk and value management (RM and VM);
- Giving contractual advice for either party in the case of dispute.

It is noted that new guidelines for the *Assessment of Professional Competence* were issued by the RICS in July 2006 (*www.rics.org.uk* [accessed 26 April 2007]).

The Construction Industry Council

The Council (representing all the professions) in their publication *Project Management Skills in the Construction Industry* (2002) identified the key skills for those operating in construction project management under the following headings:

Strategic: strategic planning/VM/RM/quality management;
Project control: the project-control cycle/developing a schedule/monitoring/managing change/ action planning/client/project interface/information management;
Technical: design management/estimating/value engineering (VE)/modelling and testing/ configuration management;
Commercial: business case/marketing and sales/financial management/procurement/legal issues;
Organization and people: organization structure/selection of a project team/people issues.

The CIOB Code of Practice for Project Management for Construction and Development

This is a significant source of knowledge with a third edition published in 2002. This CIOB initiative involved the formation of a multi-institute task force; the 1991 first edition was referred to by Sir Michael Latham in his report *Constructing the Team* (1994) as an example of cooperation between the professions.

The Association for Project Management's Body of Knowledge, 5th edition (APM, 2006)

This generic guideline for project management identifies the behavioural characteristics, the knowledge areas and competencies that are desirable for professionals in project management. It is the UK vision of generic project management within guidelines set by the International Project Management Association.

APM Body of Knowledge: key topics are listed below

Section 1 – project management in context: project management, programme management, portfolio management, project context, project sponsorship, project office;

Section 2 – planning the strategy: project success and benefits management, stakeholders management, VM, project management plan, project RM, project quality management, health and safety and environmental management;

Section 3 – executing the strategy: scope management, scheduling, resource management, budgeting and cost management, change control, earned VM, information management and reporting, issue management;

Section 4 – techniques: requirements management, development, estimating, technology management, VE, modelling and testing, configuration management;

Section 5 – business and commercial: business case, marketing and sales, project finance and funding, procurement, legal awareness;

Section 6 – organizational and governance: project life cycles, concept, definition, implementation, hand-over and close out, project reviews, organizational structure, organization roles, methods and procedures, governance of project management;

Section 7 – people and the profession: communication, teamwork, leadership, conflict management, negotiation, human resource management, behaviour characteristics, learning and development, professionalism and ethics.

1.5   Learning from case studies

Some projects become iconic, reflecting the challenge of the new and the time and place. The Sydney Opera House in Australia became the symbol for the Millennium Olympics in 2000 and somehow reflected the healthy swagger of the emerging continent. The competition for the project was won in 1957 by Danish architect Jorn Utzon, whose first design according to the members of jury was hardly more than *a few splendid line drawings*. This comment could also have been made many years later in connection with Enric Miralles' first submission on the Scottish Parliament building.

The billowing concrete sailed roof had never been built before. It should therefore have been no surprise that the A\$7 million project escalated to over A\$100 million and the planned construction period of 5 years was finally extended to 14 years (1959–1973). The architect was put under so much pressure over the escalating costs that he left the project half-way through after which his designs were modified. As a result, the building is perfect for rock concerts but not suitable for staging classical full-scale operas (Reichold, K. and Graf, B., 1999).

Another iconic building is the *Centre Pompidou in Paris*, a building famous for being *inside-out* with all the structural frame, service ducts and escalators being on the outside allowing a flexible floor space within. The audacious steel and glass National Centre for Art and Culture was designed by two young unknown architects Renzo Piano and Richard Rogers, to last – as George Pompidou reminded architects – for four or five centuries.

After opening in 1977 the centre rapidly became a huge success, with more than 7 million visitors a year making it the most popular tourist destination in Paris. After 20 years of use the building was showing its age, including rusting on the structural frame, and was in need of a major renovation. In October 1997 the whole Centre closed, reopening in January 2000, which allowed for not only the refurbishment but also improvements to the internal layout at an estimated cost of US$100 million (Poderas, J., 2002).

Significantly, the UK Government now requires the issue of whole-life costing (WLC) to be considered with project evaluation. It is interesting to speculate as to whether the Pompidou Centre would have passed such a test if the building were proposed to be built in the UK in 2007.

The *Millennium Dome at Greenwich* was designed to be UK's showcase to celebrate the new millennium. With a diameter of 365 metres the Dome was the largest single-roofed structure in the world with a floor area the size of 12 football pitches (Wilhide, E., 1999).

However, the project was plagued throughout by problems and bad publicity, typified by the Millennium New Year's Celebration, when the great and the good were left stranded for hours at Stratford Station. Open to the public for one year, the Millennium Dome was closed to visitors on 31 December 2000 and has remained unused apart from the Ministry of Sound New Year's Eve dance parties.

The £758 million project was seen as a major part in the regeneration of the East London being built on 120 hectares of contaminated wasteland. The project was initiated by the Conservatives and was taken over in 1997 by Tony Blair's new Labour Government as an innovative public sector enterprise, which harnessed private sector funding (Report by the Comptroller and Auditor General, 2000).

The Millennium Dome was designed by architect Richard Rogers and Consulting Engineer Buro Happold with a joint venture of John Laing/Sir Robert McAlpine acting as Construction Managers with target-cost contracts incorporating *painshare/gainshare* clauses. What is often forgotten is that the £260 million project for the design and construction of the Dome was a highly significant and successful achievement being completed on time and within budget. The author visited the project in 1999 and was particularly impressed with the teamwork approach with everyone: client, consultants, contractors and subcontractors working in same open-plan office.

However, the design and organization of the interior exhibits was not of the same order and some saw the result as a disjointed assemblage of thinly veiled, corporate-sponsored promotions and lacklustre museum exhibits sub-divided into 14 so-called zones. The centre-stage show with an acrobatic cast of 160 was enjoyed by most; indeed, over 80% of visitors said that they had enjoyed the whole experience.

The project was largely reported by the press to have been a flop: badly thought out, badly executed and leaving the government with the embarrassing question of what to do with it afterwards. During 2000, the organizers repeatedly asked for and received more cash from the government. Part of the problem was that the financial predictions were based on an unrealis-tically high forecast of visitor numbers at 12 million; in the event, there were only 6.5 million.

The Dome has been refurbished into a soundproof 22,000-seater sports and entertainment complex and was reopened in June 2007. In 2009 it will host the World Gymnastics Championships and in 2012 will become an Olympic venue.

The London Underground's *Jubilee Line Extension* (JLE) was at the time one of Europe's largest infrastructure projects. The JLE comprised more than 30 major contracts linking the UK Government's parliamentary HQ at Westminster to London's emerging financial centre at Canary Wharf and beyond to Stratford. The entire project comprised a 10-mile underground extension with 11 stations, 6 completely new and 5 substantially enlarged or rebuilt, and 4 crossings under the River Thames.

It posed a huge management challenge, creating magnificent station architecture, but became infamous for overrunning on both time and budget.

The overall construction period for the project was 72 months, compared to the original 53 months, while the out-turn estimated cost for the project rose from £2.1 million in October 1993 to £3.5 million in December 1999. The project was completed in December 1999, just in time for the Millennium celebrations in the Dome at Greenwich, significantly after Bechtel had taken over the management of the project in September 1998.

The JLE *Conditions of Contract* was a hybrid of the ICE 5th edition and the FIDIC form, modified by the Hong Kong Mass Transit Railway Corporation (HKMTRC) and Singapore Mass Rapid Transit. The civil engineering contracts were based on bills of quantities subject to admeasurement.

Each contract contained an Interim Payment Schedule (IPS), which was based on a series of defined milestones within four main cost centres; the idea being that payment was only made if the defined milestone was achieved. A fundamental aim of this approach was to motivate contractors to achieve progress while avoiding the need to base monthly payments on measured works. The system had worked well in Hong Kong and in the early stages of the JLE.

Unfortunately, due to the changing requirements of the JLE scheme in the early stages of the project and the tight timescale, the working drawings issued at contract award remained incomplete. This resulted in extensive changes to the programmes causing delay and disruption, extensions of time and acceleration measures, and the milestones had to be continuously revised. Indeed, many contractors commented that the whole system fell in disrepute. The Major Projects Association (MPA) observed that the project culture was too adversarial and inflexible (as were the contracts used) (MPA, 2000).

The lessons to be learned from this project are best summed up by a civil contractor:

Build your delivery strategy around collaborative working, build into your forms of contract risk and reward share and get the whole team/teams together earlier. Be decisive about objectives and keep the management team as lean as possible.

(Mitchell, R., 2003)

The Scottish Parliament Building in Edinburgh is a classic tale worthy of further investigation. As anyone will know who has experience of major projects, what happens in practice is sometimes very different to what is envisaged in theory.

Initially conceived by the Barcelona architect Enric Miralles and completed 20 months late with an initial budget of £40 million and a final cost of £430 million, the building has been subject to much criticism, particularly from the taxpayers in Scotland. Lord Frasier in his report entitled *The Holyrood Inquiry* (2004) identified many concerns including the following:

- Inadequacy of the original budget, which at the feasibility stage was no more than indicative;
- Failure to identify relative significance of cost in the time/cost/quality triangle;

- Insufficient inquiry on proposed joint venture between architects EMBT (Barcelona) and RMJM Ltd (Edinburgh);
- Inadequacy of the brief;
- Failure of client to appreciate disadvantages of the construction management approach – total risk lies with client;
- Inadequate management of risk;
- Lack of involvement of key stakeholders, for example Ministers, in procurement selection and approval of revised budget;
- Lack of audit trail in selection of construction manager (Bovis);
- Failure of client to take up Bovis's parent company guarantee;
- Lack of construction experience of client's project sponsor;
- Lack of knowledge of EU procurement rules;
- Insufficient time in the programme for the planning and design phase;
- Poor level of communication between the key players;
- Inadequate reporting to Ministers of potential risks identified by cost consultant (DL&E);
- Failure to appreciate impact of complex design, such as the chamber roof;
- Inadequacy of the cost plan, designs were developed without the cost plan;
- Failure to finalize the draft project execution plan – a key document in the control process;
- Disregard by architect of constraints of the brief and budget;
- Failure of the VM exercise;
- Ineffective monitoring systems by the project team demonstrating a lack of control over the whole process.

One comment by Lord Fraser seems to sum up the situation:

The programme was propelled by the client obsession with early completion. It appears not to have been completely grasped throughout the project that if the quality and unique complexity of the building was of overriding importance, the programme and the timing of completion would be affected significantly and extra cost inevitably occurred.

The Scottish Parliament Building was a unique project and has provided an unprecedented amount of information via the Spenceley Report (2000), two Auditor General Reports (2000 and 2004) and Lord Fraser's *Holyrood Inquiry* (2004) together with access to the actual correspondence between the parties (via the *Holyrood Inquiry* website www.holyroodinquiry.org [accessed 18 April 2007]).

In 1998 the report, *The Government Client Improvement Study*, produced by the University of Bath for the HM Treasury Procurement Group, identified that, in practice, construction project managers often resort to *crisis management*, meaning that problems are resolved through reaction rather than prevention. This report also identified that the significant lack of management effectiveness in the public sector was due to the failure of project sponsors, who are viewed as administrators rather than leaders. The report recommended that the project sponsor's role should be developed and supported by greater empowerment, training and the development of a culture that promotes and rewards openness in the way in which projects are procured, so that all can learn from good and bad experiences (*Agile Construction Initiative*, 1998).

1.6 Learning from project failures

Identifying whether a project is a success or a failure can change with time. The Sydney Opera House was clearly a failure on the criteria of budget/time/function, yet the building later became the iconic image for the nation for the 2000 Sydney Olympics.

Many of the 1970s UK North Sea oil projects went way over the budget, yet following the subsequent surge in oil prices were clearly successful. The Thames Barrier at Greenwich was a project plagued by poor industrial relations finishing three years late in 1984 and 10 times the original budget, yet when the barrier was later used, the innovatively designed project was successful and London was saved from flooding. Today the Barrier is raised six times per year compared to once in six years as originally anticipated.

It is important that we understand the causes of project failures and apply the lessons learned. The Major Project Association (MPA, 2003) identifies the major reasons for project failure as follows:

- Poor project definition;
- Unclear objectives;
- Unrealistic targets;
- Inadequate risk evaluation;
- Client inexperience;
- Poor forecasting on demand;
- Lack of effective sponsor and strong leadership;
- Poor communication and lack of openness;
- Inadequate stakeholder management;
- Management focus wrongly targeted at the back end rather than at the front end of the project.

These are indeed important observations that ring true in the case studies quoted throughout this book.

1.7   Relevant observation

In 1913 the *Builder* magazine criticized the standard form of contract and concluded, 'It does seem as though there is an appalling waste of money and effort in the tendering system . . . and it would be to the benefit of the building owners if this waste was curtailed.' The same article went on to argue:

> The competitive system of tendering . . . divorces the interest of the contractor from that of the building owner. If some system could be devised whereby the contractor was paid in a way which would make him . . . stand in the shoes of the building owner, it would confer on the latter many of the advantages of doing the work himself.
>
> (Quoted in Cooper, 2000)

Nearly one hundred years later we can confidently state that some progress has been made in developing procurement systems that enable contractors to stand in the shoes of the building owner. However, there is still much work to be done in identifying and disseminating best practice.

1.8   Conclusion

This chapter has set the scene and identified the role of the construction commercial manager within the wider discipline of project management. The role of the commercial manager has developed embracing many aspects of project management including the following:

- Strategic: strategic planning, VM and RM, WLC;
- Control: project-control cycle, developing a schedule, controlling the cost;

- Technical: tendering procedures, contractors' estimating;
- Commercial: financial management of the pre-tender, tender and post-tender stages, procurement, managing change, managing claims, legal awareness and contracts.

The aim of this book is to embrace the subject of construction-cost management as identified within sections 2, 3, 4 and 5 of the APM's Body of Knowledge, the CIC's Construction Project Management Skills framework and the RICS's APC/ATC guidelines for construction surveyors. The linking of these sections acknowledges the reality of the role of the QS/commercial manager, increasingly embracing construction project management.

A brief analysis of two important forms of contract, the NEC3 and FIDIC's 1999 *Red Book*, representing best practice in the UK and overseas is included. The final chapter describes the Heathrow Terminal 5 project, which has embraced the lean construction philosophy demanded by Sir John Egan. Particular attention is paid to the recommendations made in the key reports and from the National Audit Office. This text should therefore prove to be of interest to built environment and project management undergraduate/postgraduate students and practitioners alike.

Commercial and contractual management is not easy; traditionally it demanded experienced, dedicated personnel with an understanding of the construction technology and an in-depth knowledge of measurement and estimating, variations and claims and contract procedures. The new strategies now demand professionals with a wider knowledge of procurement strategies and project management – particularly planning and control systems, corporate governance, strategic positioning, organizational behaviour, supply-chain management and the management of change.

However, it is not the strategy or the wording of the contract that ensures success or failure, rather it is the attitude of the people involved. A genuine team spirit must be created with all team members having a *can do, will do* attitude (MPA, 2001).

Now it's time for you to answer a few questions! Best of luck.

1.9   Questions

1.   Identify the changing roles of the commercial manager within the UK construction industry.
2.   What are the qualities required of a successful commercial manager within the construction sector?
3.   How do clients control their investments and when?
4.   How can the contractor control the project?
5.   How can a contract contribute to effective project management?

Bibliography

Agile Construction Initiative, University of Bath (1998) 'The Government Client Improvement Study', A Report for HM Treasury Procurement Group and the Government Construction Client Panel (GCCP)

APM (2006) *APM Book of Knowledge*, 5th edition, Association for Project Management, High Wycombe, UK (ISBN:978–1–903494–13–4). Reproduced with permission

Auditor General for Scotland (2000) *The New Scottish Parliament Building: An Examination of the Management of the Holyrood Project*, Audit Scotland

Auditor General for Scotland (2004) *Management of the Holyrood Building Project*, Audit Scotland

CIOB (2002) *Code of Practice for Project Management for Construction and Development*, 3rd edition, CIOB

Construction Industry Council (2002) *Construction Project Management Skills*, CIC

Cooper, P. (2000) *Building Relationships. The History of Bovis 1885–2000*, Cassell & Co

Egan, J. (1998) *Rethinking Construction: Report of the Construction Task Force on the Scope for Improving Quality and Efficiency of UK Construction*, Department of the Environment, Transport and the Regions (DETR)

FIDIC (1999) *Conditions of Contract for Building and Engineering Works Designed by the Employer*, 1st edition, Federation Internationale des Ingenieurs-Conseils

Fraser Rt Hon. Lord of Carmyllie (2004) *The Holyrood Inquiry*, SS Paper No. 205, Session 2, Astron, Edinburgh

ICE (1973) *ICE Conditions of Contract*, 5th edition, Thomas Telford

Latham, M. (1994) *Constructing the Team, Joint Review of Procurement and Contractual Arrangements in the United Kingdom Construction Industry*, Final Report, HMSO

Major Projects Association (2000) 'The Jubilee Line Extension', Seminar held at the ICE, London, 17 November

Major Projects Association (2001) 'Hong Kong International Airport', Seminar held the ICE, London, 23 April

Major Projects Association (2003) 'Learning from Project Failures', Seminar held at the Royal College of Pathologists, London, 13 November

Mitchell, R. (2003) *Jubilee Line Extension: From Concept to Completion*, Thomas Telford

Murray, M. and Langford, D. (eds) (2003) *Construction Reports 1944–98*, Blackwell Publishing

National Audit Office (2005) *Improving Public Services through Better Construction*, Report by the Comptroller and Auditor General, HC364, Session 2004–2005: 15 March, HMSO

NEC (2005) *NEC3 Engineering and Construction Contract*, Thomas Telford

Poderas, J. (2002) *Centre Georges Pompidou Paris*, Prestel

Reichold, K. and Graf, B. (1999) *Buildings That Changed the World*, Prestel

Report by the Comptroller and Auditor General (2000) *The Millennium Dome*, HC936, Session 1999–2000: 9 November, The Stationery Office

RICS (2002) *APC/ATC Requirements and Competencies*, July, 1st edition, RICS

Spenceley, J. (2000) *Scottish Parliamentary Corporate Body Report on the Holyrood Project*, SP Paper 99, Session 1, The Stationery Office

Wilhide, E. (1999) *The Millennium Dome*, Harper Collins Illustrated

2 Reports and recommendations

2.1 Introduction

The past decade or so has seen the publication of many significant reports relevant to project management of the built environment (embracing both building and civil engineering), including significantly Latham (1994) and Egan (1998) and those issued by the UK Government's Office of Government Commerce (part of HM Treasury) and the National Audit Office.

These major studies have highlighted the inefficiencies of traditional methods of procuring and managing major projects – in particular, the fallacy of awarding contracts solely on the basis of the lowest price bid, only to see the final price for the work increase significantly through contract variations, with projects often completed late. Indeed, this was often the traditional ploy on major works – submit a low bid in the anticipation of making a profit on the variations and claims.

Experience has shown that acceptance of the lowest price bid does not provide value for money in either the final cost of construction or through-life and operational costs. Relations between the construction industry and government departments have also been typically characterized by conflict and distrust which have contributed to poor performance.

> Estimates of the cost of these inefficient practices are inevitably broad brush, but studies have identified the potential for major savings – 30 per cent in the cost of construction. Specifically by industry adopting a more collaborative approach strongly founded on a competitive process with appropriate risk sharing in which value for money is obtained by all parties through a clear understanding of the project's requirements, transparency as to costs and profits, underpinned by clear understood rights and obligations, and appropriate incentives. More attention to design and early involvement of the whole construction team could also improve the operational efficiency of the completed buildings resulting in potentially greater savings over the whole life of the building.
>
> (NAO, *Modernising Construction*, 2001, p. 4)

It is appropriate to review the recommendations contained within the major reports; the list is not exhaustive but should contain the key criteria for change and enable identification of recommended best practice.

Unlike previous UK Government reports on the construction industry, which were basically ignored, the Latham report *Constructing the Team* (1994) and the Egan report *Rethinking*

Construction (1998) have had a profound impact on the UK construction industry. These reports, together with Egan's follow-up report *Accelerating Change* (2002) have challenged the industry to throw off the old adversarial practices and reinvent itself in order to become world-class.

The UK Government as a best-practice client has instigated major changes in procurement and project management practice. Case studies indicate that the new approaches are having a significant impact on increasing the client's certainty of outcome and value for money.

In the mid-1990s the sharp reversal in fortunes and prospects of most of those in the construction industry, professionals, contractors and suppliers, prompted a radical review of the established practices and procedures. More significantly, it created a willingness to work together to consider changes and how best they be implemented for the benefit of the industry and the clients on whom it depends.

It is important that all those engaged in the construction process understand these changes as they come into effect through parliamentary statute, new forms of contract and codes of practice. All should be conscious of the changed responsibilities and liabilities that will arise and the opportunities as well as risks that they provide for their business.

2.2 The Latham report, *Constructing the Team* (1994)

The terms of reference were to consider the following.

Current procurement and contractual arrangements; and current roles, responsibilities and performance of the participants including the client with regard to:

- the processes by which clients' requirements are established and presented;
- methods of procurement;
- responsibility for the production, management and development of design;
- organization and management of the construction process;
- contractual issues and methods of dispute resolution.

The report makes 30 major points to which the industry has responded well. Perhaps most significant was the establishment of the Construction Industry Board (CIB) to which all parties in the industry contributed and supported. A number of working groups were set up under the aegis of the CIB to find ways to implement the report's recommendations.

Main conclusions and recommendations

Clients

Government and private sector:

- set up a new construction clients' forum;
- be best-practice clients;
- publish a Construction Procurement Strategy Code of Practice;
- promote a mechanism for selecting consultants on quality as well as price.

Industry:

- adopt target of 30% real cost reduction by the year 2000;
- improve tendering arrangements/registration (with government);
- draw up a code of practice for selecting subcontractors;

- implement the recent reports on training and the education of professionals;
- improve public image;
- produce coordinated equal opportunities action plan.

Contracts:

- develop standard contract documentation based on a set of principles (including independent adjudication, pre-pricing of variations and Trust Accounts for payments);
- recommendations for increased use of the New Engineering Contract;
- produce a complete standard family of interlocking contract documentation;
- contract committees – restructuring.

Legislation:

- introduce legislation against unfair contracts;
- introduce legislation to underpin adjudication and Trust Account proposals;
- implement Department of Environment working-party proposals on liability legislation;
- introduce mandatory latent-defects insurance.

Sir Michael Latham also identified what a modern contract should contain, which is particularly relevant when considering the contract between the main contractor and the subcontractor.

Panel 2.1 A modern contract

The most effective form of contract in modern conditions should contain the following:

1. A specific duty for all parties to deal fairly with each other, and with their subcontractors, specialists and suppliers, in an atmosphere of mutual cooperation.
2. Firm duties of teamwork, with shared financial motivation to pursue those objectives. These should involve a general presumption to achieve 'win–win' solutions to problems which may arise during the course of the project.
3. A wholly integrated package of documents which clearly defines the roles and duties of all involved, and which is suitable for all types of projects and for any procurement route.
4. Easily comprehensible language and with 'Guidance Notes' attached.
5. Separation of the roles of contract administrator, project or lead manager and adjudicator. The project or lead manager should be clearly defined as the client's representative.
6. A choice of allocation of risks, to be decided as appropriate to each project but then allocated to the party best able to manage, estimate and carry the risk.
7. Taking all reasonable steps to avoid changes to pre-planned works information. But, where variations do occur, they should be priced in advance, with provision for independent adjudication if agreement cannot be reached.
8. Express provisions for assessing interim payments by methods other than monthly valuation, that is, milestones, activity schedules or payment schedules. Such arrangements must also be reflected in the related subcontract documentation. The eventual aim should be to phase out the traditional system of monthly measurement or re-measurement, but meanwhile provision should still be made for it.

---

9. Clearly setting out the period within which interim payments must be made to all participants in the process, failing which they will have an automatic right to compensation, involving payment of interest at a sufficiently heavy rate to deter slow payment.
10. Provide for secure trust fund routes of payment.
11. While taking all possible steps to avoid conflict on site, providing for speedy dispute resolution if any conflict arises, by a predetermined impartial adjudicator/referee/expert.
12. Providing for incentives for exceptional performance.
13. Making provision where appropriate for advance mobilization payments (if necessary bonded) to contractors and subcontractors, including in respect of off-site prefabricated materials provided by part of the construction team.

Source: Latham, 1994.

2.3 Levene Efficiency Scrutiny (1995)

Although the Latham report clearly tried to improve performance of the UK construction industry and provide a catalyst for change, further reforms were required especially in the way government departments procured contracts. The Levene Scrutiny focused in greater detail on the role of various government departments and agencies in the procurement of construction work and how they would perform being *best-practice clients* (Efficiency Unit Cabinet Office, 1995).

The review was undertaken with two fundamental aims: to improve value for money in the procurement of public works and to improve the competitiveness of suppliers to government.

The report concluded, as Latham had already stated, that the UK construction industry was in poor shape and that the performance of the government departments was a contributing factor. The report proposed 5 action points, developed into 22 recommendations which were designed to facilitate government departments into managing their projects more effectively, and to encourage the industry to be more proactive and less adversarial.

2.4 *Construction Procurement Guidance*, HM Treasury (1996)

This series of guides were produced following the recommendations of the Latham report (1994) and the Efficiency Scrutiny of Government Construction Procurement (1995). The guidance provided best-practice advice at a strategic level and covered the client's role in the procurement process; they were specifically aimed at encouraging a change in culture.

The reports made up a family of documents comprising:

No. 1 Essential requirements for construction procurement

Set out roles and responsibilities of the investment decision-maker, the project owner and project sponsor and the training they required.

No. 2 Value for money (VFM) in construction procurement

Set out a VFM framework (a structured list of activities undertaken in a project including *approval gateways*, risk and value management techniques and control procedures).

Best VFM is the optimum combination of whole life cost and quality to meet the customers' requirement.

The National Audit Office (NAO) does not consider that achieving VFM means accepting the lowest bid – they have not criticized a project on these grounds when other considerations were more important. Neither do they wish to stifle innovation through rigid adherence to mechanistic procedures.

No. 3 Appointment of consultants and contractors

Set out the consultancy roles and responsibilities, details of the appointment process and the structure of the project team.

No. 4 Team working, partnering and incentives

Embraced the concept of teamwork declaring that 'Teamwork should be a core requirement for every element of a major project and partnering should be adopted as far as possible on all new and existing contracts. Incentives should be included to provide benefits to clients.'

No. 5 Procurement strategies

Recommended the following procurement strategies: Public–Private Partnerships (PPP), design and construct (and where appropriate maintain and operate), prime contracting and framework agreements; traditional forms of construction procurement should only be used where there is a very clear case that they will deliver better value for money than other procurement routes in terms of whole-life costs and overall performance.

No. 6 Financial aspects of projects

Provided information on preparing budget estimates and dealing with risk allowances.

No. 7 Whole-life costs

This guide identified that the primary purpose of whole-life costs is to provide the investment decision maker with the information necessary to make the best decisions in terms of project strategy and procurement route.

No. 8 Project evaluation and feedback

Identified that project evaluation includes three elements:

- Formal reviews at project gateways (including post-occupancy evaluation);
- Less formal ongoing evaluation and reporting (particularly during the development and construction stages);
- Evaluation and reporting of specific activities.

The fundamental part of project evaluation and feedback is to make sure that lessons learned from one project are transferred effectively to other projects.

No. 9 Benchmarking

The guide considered that the primary purpose of benchmark is to improve the performance of the organization. Benchmarking is a tool that allows organizations to help themselves. It is an essential part of continuous improvement and is a continuous and long-term process and not a one-off instant solution.

2.5 Construction Industry Board (CIB) working groups (1996–1997)

Following Sir Michael Latham's report *Constructing the Team* the CIB representing the professions and consultants, main contractors, subcontractors, clients, materials suppliers and the government produced reports from 12 working groups as follows:

WG1	Briefing the team
WG2	Constructing success
WG3	Code of practice for the selection of subcontractors
WG4	Selecting consultants for the team: balancing quality and price
WG4	Framework for a national register for contractors
WG5	Framework for a national register for consultants
WG6	Training the team
WG7	Constructing a better image
WG8	Tomorrow's team: women and men in construction
WG9	Educating the professional team
WG10	Liability law and latent defects insurance
WG11	Towards a 30% productivity improvement in construction
WG12	Partnering in the team.

The document WG1 *Briefing the Team* contains a checklist for construction productivity – this is essentially the solution to Latham's challenge for the industry to make a 30% real cost reduction. With the benefit of hindsight, see if you can think of any other relevant items which are not included in the WG1 report.

- Change the industry culture;
- Introduce clear, concise and comprehensive standards of briefing;
- Ensure that design and construction processes work as one;
- Foster teamwork and partnership;
- Rationalize project structures;
- Establish industry standards for information technology;
- Make quality the main requirement of all elements of the design and construction process;
- Improve the understanding and effective application of risk-management techniques;
- Health and safety should be part of the cost–benefit analysis;
- Develop standard products, components and processes;
- Prefabrication and preassembly should be part of design considerations;
- Improve designers' knowledge and understanding of the performance of components and materials;
- Designers need urgently to embrace new technologies;
- Life cycles, and whole-life costs of buildings and their fittings must be a principal part of design and maintenance considerations;
- Quality and value must not be ignored in the pursuit of the lowest price;
- The management experience of buildings and projects, and the associated costs, should be constantly fed back to, and adopted by, designers in new designs;
- Benchmarking must be used to measure improvements of practice and productivity;
- Shared construction experience must be given to trainees during their education;

- Focus on research and innovation, integrate current research projects, improve information flow; invest in implementation;
- Establish public relations channel, focus on productivity gains, highlight successes.

2.6   The Egan report *Rethinking Construction* (1998)

Rethinking Construction is the name of the report produced by Sir John Egan's Construction Task Force. The report commissioned by John Prescott, the Deputy Prime Minister, was published in July 1998. The central message of *Rethinking Construction* is that through the application of best practices, the industry and its clients can collectively act to improve their performance.

The *Rethinking Construction* report proposed the creation of a *movement for change* which would be a dynamic, inspirational, non-institutionalized need for radical continuous improvement within the construction industry. The report led to further action to facilitate cultural change; with particular emphasis on the need for involvement of the whole of the supply chain. Another such change was the launch of the Movement for Innovation (M4i) in November 1998, which since 2004 has been part of *Constructing Excellence*.

The report also encouraged the recognition that the industry can and indeed must do much better. This led to M4i capturing 180 demonstration projects submitted by clients and contractors, which exemplified some of the innovations advocated in Sir John Egan's report. Many of the demonstration projects did exceed Sir John Egan's targets in productivity, profits, defects and reduced accidents.

The report can be summarized within the 5:4:7 mantra: five 'drivers' that needed to be in place to secure improvement in construction; four processes that had to be significantly enhanced; seven quantified improvement targets.

Drivers for change

1. Committed leadership
2. Focus on the customer
3. Product team integration
4. Quality-driven agenda
5. Commitment to people.

Improving the process

1. Product development
2. Partnering the supply chain
3. Product implementation
4. Production of components.

Targets for improvement (annual)

1. Capital cost   − 10%
2. Construction time   − 10%
3. Predictability   +20%
4. Defects   − 20%
5. Accidents   − 20%
6. Productivity   +10%
7. Turnover and profits +10%

To enable the construction industry to achieve the targets, radical changes were identified within the Egan report. One such change was the replacement of traditional contract strategies with integrated supply-chain-led strategies, such as design and build, alongside long-term partnering relationships based on clear measurement of performance and sustained improvements in quality and efficiency, which continued the theme from the earlier Latham report.

Key recommendations within the Egan report were summarized by Bennett and Baird (2001) as follows:

- The industry and its major customers need to rethink construction so as to match the performance of best consumer-led manufacturing and service industries.
- Integrated processes and teams should be introduced as a key driver for change.
- The industry should organize its works so that it offers customers brand-named products, which they can trust to provide reliably good value.
- The industry should work through long-term relationships using partnering, which aims at continuous improvements in performance.
- Benefits from improved performance should be shared on an open, fair basis so that everyone has real motivation to search for better answers.
- Project teams should include design, manufacturing and construction skills from day one so that all aspects of the processes are properly considered.
- Decisions should be guided by feedback from the experience of completed projects so that the industry is able to produce new answers that provide even better value for the customer.
- Standard products should be used in designs wherever possible because they are cheaper and, in the hands of talented designers, can provide buildings that are aesthetically exciting.
- Continuous improvements in performance should be driven by measured targets because they are more effective than using competitive tenders.
- The industry should end its reliance on formal conditions of contract because in soundly based relationships in which the parties recognize the mutual interdependence, contracts add significantly to the cost of projects and add no value to the customer.

One of the more controversial comments made within the Egan report included the following:

The Task Force wishes to see: an end to the reliance on contracts. Effective partnering does not rest on contracts. Contracts can add significantly to the cost of a project and often add no value for the client. If the relationship between a constructor and employer is soundly based and the parties recognize their mutual interdependence, then formal contract documents should gradually become obsolete. The construction industry may find this revolutionary. So did the motor industry but we have seen non-contractually based relationships between Nissan and its 130 principal suppliers and we know they work.

In reality, this may be a step too far for many within the construction industry.

The targets set by *Rethinking Construction* have been met by several major construction clients. However, after achieving the necessary cost and time reduction for three years in running without the anticipated increase in turnover and profits, there comes a time when contactors begin to wonder whether it has been worth the effort (comment from commercial manager of major contractor in 2004).

The UK Government policies have now increased the need for all public sector clients to fully implement the principles of *Rethinking Construction* which are now firmly established and recognized as best practice.

2.7 *Modernising Construction*, National Audit Office (2001)

This report identified how the procurement and delivery of construction projects in the UK could be improved. Its recommendations were made to four key groups: the Department of the Environment, Transport and the Regions, the Office of Government Commerce, line departments commissioning construction projects and the construction industry itself. The main recommendations are summarized as follows (see also Fig. 2.1):

The Department of the Environment, Transport and the Regions

- Provide more coordinated direction to initiatives to promote better performance by the construction industry;
- Use its influence as a Member for Innovation Board to ensure that demonstration projects are truly innovative;
- Develop more sophisticated performance measures, for example, indicators needed to measure:
    a. The operational – through-life running costs of the completed building;
    b. The cost-effectiveness of the construction process;
    c. Quality of completed construction;
    d. Health and safety indicators.

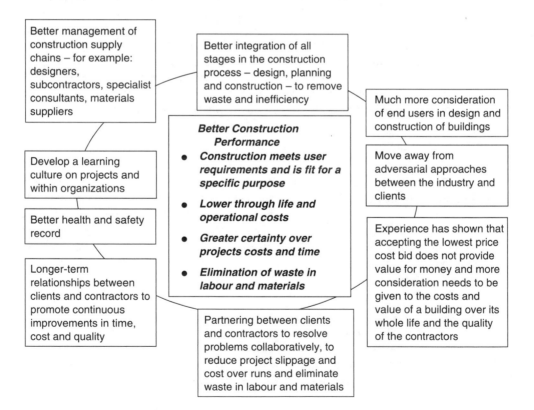

2.1 Better construction performance what is needed? (source: NAO analysis of Latham, Levene and Egan reports, Copyright National Audit office, reproduced with kind permission, reported in *Modernising Construction*, 2001).

The Office of Government Commerce

- Disseminate good practice more widely;

Line departments

- Actively measure improvements in construction performance;
- Train more staff to be effective construction clients.

The construction industry

Make greater use of innovation to improve public sector construction.

2.8   The second Egan report, *Accelerating Change* (2002)

Four years on from the Egan report the Strategic Forum for Construction produced a follow-up report titled *Accelerating Change*. This report tackled some of the barriers to progress against the Egan report's targets and identified ways of accelerating that change.

The vision of the follow-up report was for the UK construction industry to realize maximum value for all clients, end-users and stakeholders and exceed their expectations through consistent delivery of world-class products. The vision is being taken forward by the Strategic Forum for Construction (SFfC), a cross-industry/government body.

The following are the four key areas on which the SFfC focused:

1. Client engagement
2. Integrating teams and supply chains
3. People issues
4. Enhancing the value of the product.

   Six headline targets were identified by the SFfC:

Target 1: 20% of construction projects (by value) should be undertaken by integrated teams and supply chains by end of 2004, rising to 50% by end of 2007;

Target 2: 20% of clients to embrace the principles of the Clients' Charter by 2004. Target to increase to 50% by 2007;

Target 3: by 2006, 300,000 qualified people to be recruited and trained in the industry;

Target 4: by 2007, a 50% increase in applications to built environment higher- and further-education courses, and by 2010 an increase in the annual rate of apprentice completions to 13,500;

Target 5: by 2010, a fully trained, qualified and competent workforce on all projects;

Target 6: by the end of 2004, 500 projects to have used the design quality indicators (DQIs). By the end of 2007, 60% of all publicly funded Private Finance Initiative (PFI) projects (having a value in excess of £1 million) to use DQIs and 20% of all projects (having a value in excess of £1 million to use DQIs.

In essence the *Accelerating Change* reinforced the challenges set out in *Rethinking Construction* calling for a most ambitious year-on-year improvement.

2.9 *Achieving Excellence in Construction Procurement Guides,*
Office of Government Commerce (2003)

Achieving Excellence was launched in March 1999 to improve the performance of central government departments and other public bodies, following major failures in time and cost overruns. It aimed to provide a step-by-step change in construction procurement performance and value for money achieved by government on construction projects, including maintenance and refurbishment.

The key thrust of *Achieving Excellence* is the delivery of value for money. This is not the lowest cost but the optimum combination of whole-life cost and quality to meet the users' requirement.

The *Achieving Excellence* suite of procurement guides replaced the 1996 *Construction Procurement Guidance* series. The new series reflected developments in construction procurement over recent years and built upon government departments' experience of implementing the *Achieving Excellence in Construction* initiative.

The suite consists of three core and eight supporting documents together with two high-level documents. Electronic versions have hyperlinks across the set and to related products, such as the OGC *Successful Delivery Toolkit* and external websites. The significant OGC *Gateway Process* model is described in detail in the *Achieving Excellence Guide 3 – Project Procurement Lifecycle*.

2.10 *Improving Public Services through Better Construction,*
National Audit Office (2005)

Part 3 of the report and the supporting case studies set out examples of good practice which have enabled organizations in both the public and private sectors to improve their construction delivery. The good practices have allowed the completed projects to be delivered on time and to cost and have helped improve the quality of the final built asset. In addition, the report recommended that government departments initiate the following:

1. Create more certainty in the market, with longer-term funding and programme planning;
2. Strengthen their leadership of construction programmes and projects and put in place strategies for developing construction project management capabilities;
3. Engage fully with the Gateway process and obtain independent advice and challenge at the concept and business-case stages when considering potential construction projects;
4. Consider the development of a sustainability action plan to cover all aspects of their construction activity;
5. Make decisions about construction projects based on sustainable whole-life value;
6. Make more transparent to suppliers the criteria for tender evaluation and make the most of their funding and purchasing power to influence suppliers' behaviour;
7. Keep competitive tension in framework and partnering arrangements to provide greater assurance that construction costs represent fair value, and improve the effectiveness of contract strategies to manage better risk and maximize the opportunities for improved performance;
8. Encourage collaborative working through collaborative forms of contract and fair payment practices and seek opportunities to pursue the case for project-wide insurance where appropriate and in agreement with their suppliers;
9. Evaluate the post-completion and occupancy performance of projects in terms of the *Achieving Excellence* strategic targets, whole-life value, including the financial performance and the delivery of better services and sustainable development, and embed the lessons in future activity;
10. Relevant departments should consider developing quantifiable cross-government strategic targets focused on sustainable construction.

Additional recommendations were made to the OGC advising them on the leadership and support which they should provide to all public sector organizations. The report also included a useful self-assessment tool in the form of a *maturity grid* enabling public sector clients to assess their readiness and capability to tackle construction projects and to target areas for improvement.

2.11   Conclusion

These reports have identified a substantial contribution to the construction of best-practice body of knowledge; they have identified the following key issues in order to improve the project management process:

- Leadership and commitment from the client's senior representative;
- Involvement of the key stakeholders throughout the project;
- Roles and responsibilities clearly understood by everyone involved in the project, with clear communication lines;
- An integrated project team consisting of client, designers, constructors and specialist suppliers, with input from facilities managers/operators;
- An integrated procurement process in which design, construction, operation and maintenance are considered as a whole;
- Design that takes account of functionality, appropriate build quality and impact on the environment;
- Commitment to excellence in health and safety performance;
- Risk and value management that involves the entire project team, actively managed throughout the project;
- Award of contract on the basis of best value for money over the whole life of the facility, not lowest tender price;
- Commitment to continuous improvement;
- Commitment to best practice in sustainability.

These factors challenge the industry to produce projects which achieve *best value* with a need to understand the balance between quality and whole-life cost.

2.12   Questions

1.   How do the contracts used within your organization score against Sir Michael Latham's recommendations for a modern effective contract? Consider contracts between the employer and the contractor, and the contractor and its subcontractor.
2.   An international contractor wins a £50 million project for construction of fast-track, high-rise, city-centre tower block from a major financial institution with whom they've worked many times before. The relationship between the constructor and employer is soundly based. The contract is contained in 120 words on one page including just the name of the client and contractor, their addresses, the project name and location, the design documents, the start and completion dates, contract sum and the payment schedule.

     The client introduces changes and acceleration instructions throughout. The final cost of the project to the contractor is £60 million. The contractor argues that the contract should be valued on cost-reimbursement basis. The client's retort is that the £50 million quoted was on a guaranteed maximum-price basis.

     After three years of bitter negotiating the parties have still not settled and agree to go to arbitration. The lawyer's fees for each side is £5 million and rising!

     a.   Is this scenario possible? You decide.
     b.   Identify what is the purpose of a contract.

3. Critically compare and contrast the OGC Gateway™ Process with the RIBA Plan of Work. Source: OGC framework for construction procurement: www.ogc.gov.uk/documents/ CP0063AEGuide3.pdf (pp. 6–7, accessed 6 June 2007); RIBA Plan of Work: www.a4a.sk/ RIBA-outline.htm (accessed 6 June 2007).
4. Identify the changes to practices and procedures which were made within your own organization (either employer or main contractor, and specialist contractor or consultant) in the past decade.
5. Critically review the lessons that the construction industry can learn from other sectors (contained within *Rethinking Construction*).
7. Review one of the five M4i demonstration projects included in *Modernising Construction* (appendices 9 to 13) and make a 10-minute presentation to your peers on the key recommendations.
8. Critically review and identify the key lessons learnt in 1 of the 10 case studies in the National Audit Office Report, Case Studies *Improving Public Services through Better Construction*, 2005 and make a 10-minute presentation to your cohort (source: http:// 195.92.246.148/nhsestates/procure21/p21_content/home/documents/MkCaseStudy.pdf [accessed 23 March 2007]).

Bibliography

Bennett, J. and Baird, A. (2001) *NEC and Partnering: The Guide to Building Winning Teams*, Thomas Telford

Construction Industry Board (1996/97) *Construction Industry Board Reports*, Thomas Telford

Dixon, M. (ed.) (2000) *Project Management Body of Knowledge*, 4th edition, Association for Project Management

Efficiency Unit Cabinet Office (1995) *The Levene Scrutiny*, HMSO

Egan, J. (1998) *Rethinking Construction: The Report of the Construction Task Force to the Deputy Prime Minister, John Prescott, on the Scope for Improving the Quality and Efficiency of UK Construction*, London, Department of the Environment Transport and Regions Construction Task Force

Latham, M. (1994) *Constructing the Team: Final Report of the Government/industry Review of Procurement and Contractual Arrangements in the UK Construction Industry*, HMSO

National Audit Office (2001) 'Modernising Construction', Report by the Comptroller and Auditor General, HC87, Session 2000–2001: 11 January, London, The Stationery Office

National Audit Office (2005) *Improving Public Services through Better Construction*, The Stationery Office

National Audit Office (2005) *Improving Public Services through Better Construction: Case Studies*, The Stationery Office

OGC (2003) *Achieving Excellence in Construction Procurement Guides 01 to 11*, Office of Government Commerce

Strategic Forum for Construction (2002) 'Accelerating Change', A Report by the Strategic Forum for Construction Chaired by Sir John Egan, Rethinking Construction c/o Construction Industry Council

Useful websites

www.clientsuccess.org.uk (Confederation of Construction Clients)

www.constructingexcellence.org.uk (identifies and disseminates best practice)

www.nao.gov.uk (National Audit Office)

www.ogc.gov.uk (Office of Government Commerce)

Part II
Management of the pre-contract stage

3 Selecting the consultants and contractors

3.1 Introduction

The client is the key member of any team engaged in a construction project and will need to establish means of acting efficiently within the team to ensure the success of the project. Responsibilities of the client are both legal and contractual and include the following:

- Appointment of consultants and contractor(s);
- Health and safety under the CDM regulations;
- Defining and specifying the outcomes required from the project;
- Making appropriate decisions and giving approvals within a set timescale;
- Providing payment to contracted parties for services provided.

The client may also wish to consider at the outset of the project how many of the client's responsibilities they may wish to delegate to the consultant/construction team, for example, through the appointment of a client's representative, the delegation of design or cost decisions, or the entire transfer to others by means of a financing agreement to construct and manage the project.

> It's unwise to pay too much, but it's worse to pay too little. When you pay too much, you lose a little money – that is all. When you pay too little, you sometimes lose everything, because the thing you bought was incapable of doing the thing it was bought to do.
>
> The common law of business balance prohibits paying a little and getting a lot – it can't be done. If you deal with the lowest bidder, it is well to add something for the risk you run. And if you do that, you will have enough to pay for something better.
>
> (John Ruskin, 1860)

It is now a UK Government requirement that all public sector procurement should be on the basis of value for money (VFM) and not lowest price alone; this philosophy applies to the selection of both consultants and contractors.

The recommendations require that robust mechanisms should be developed to evaluate the quality and price, including whole-life costs, in a fair and transparent manner. Selection procedures are also required to comply with the EU procurement rules where these are applicable.

There are three separate stages in the appointment process of consultants and contractors:

Stage 1 – the initial stage

During the initial stage it is necessary to identify what the consultant or contractor should do under the contract, consideration of the selection options including open, selective or negotiated, identification of specific health and safety requirements, development of the contract requirements and, in the public sector, consideration of the EU procurement directives.

Stage 2 – the selection process

The second stage involves setting the selection and award criteria, inviting expressions of interest, developing a long list and reducing it to a short list. In the public sector this will involve advertising in the *Official Journal of the European Union* (OJEU) – formerly known as the OJEC. On major projects this will normally involve the compilation of a pre-qualification questionnaire.

The selection process will involve the following stages:

- Establishing the selection criteria;
- Developing the weightings for the selection criteria;
- Identifying, where appropriate, the thresholds for the selection criteria;
- Establishing the selection mechanism;
- Inviting expressions of interest/drawing up a long list;
- Drawing up the short list.

Stage 3 – the award process

The third stage involves interviewing and inviting tenders, evaluating tenders, negotiating and awarding the contract and finally debriefing all tenderers.

3.2   Selecting consultants

Consultants and client advisers provide the foundation on which a successful project is constructed. Depending on the client's in-house resource, selected consultants could provide the following functions:

- Design services (single organization could be responsible for all design duties, with other designers appointed as subcontractors; alternatively different organizations could be appointed for each of the key disciplines with the project manager responsible for managing and controlling them);
- Project management (including cost management);
- Value management, risk management, partnering facilitator, facilities management (may be included in project management).

The consultant's brief should describe the services that the consultant is required to carry out precisely. These are dictated by the strategy adopted and whether the services are to be provided individually or in combination.

The project sponsor must also ensure that any authority delegated to the project manager is carefully defined particularly in connection with the following:

- Order variations and make changes;
- Certify interim payments;
- Grant extensions of time;

- Settle claims;
- Agree final accounts.

Developing the model for selecting consultants

Establishing VFM has as much to do with the quality of goods and services as with their price. But there must be a sound basis for evaluation and judgement.

Sir Michael Latham in his 1994 report *Constructing the Team* stated that 'professional consultants should be selected on a basis which properly recognizes quality as well as price.' Working group 4 of the Construction Industry Board (CIB) was established to choose and endorse a quality, price-assessment mechanism for appointing professionals – including architects, engineers, surveyors and project managers.

The principal features of the quality/price mechanism recommended in this report are summarized as follows:

1. The quality/price mechanism should be established by a formally constituted and fully accountable tender board before tenders are invited, and all tender documentation should be designed to ensure that appropriate responses are received to which the mechanism can be applied.
2. A quality/price ratio must be agreed at the outset, representing the percentage weightings to be given to quality and price. The more complex the project, and the greater the degree of innovation and flexibility likely to be required from the consultants, the higher the ratio should be. Indicative ratios suggested for various types of projects are

Type of project	Indicative quality/price ratio
Feasibility studies and investigations	85/15
Innovative projects	80/20
Complex projects	70/30
Straightforward projects	50/50
Repeat projects	20/80

3. Quality criteria (see also Table 3.1) should be grouped under four main headings and weighted. Recommended headings and suggested weightings are

Quality criteria	Suggested weighting range (%)
Practice or company	20–30
Project organization	15–25
Key project personnel	30–40
Project execution	20–30

4. A quality threshold needs to be established (e.g. 65 out of 100). Tenders must achieve this minimum quality score before final interviews are held and prices considered.
5. Submitted tenders are assessed for quality by marking each of the 4 quality criteria out of 100, multiplying each mark by the respective weighting percentage and then adding them together to give a total score out of 100.
6. Consultants passing the quality threshold (ideally only two to three) are then interviewed, their quality scores are reviewed and their prices examined and marked. The lowest compliant bid scores 100 and others score 100 minus the percentage figure above the lowest prices (e.g. a bid 25% above the lowest scores 75).

Table 3.1 Project-specific criteria.

Generic quality criteria (marked out of 100)	Key aspects	Suggested weighting range (%)
Practice or company	Organization Financial status Professional indemnity insurance Quality assurance or equivalent system Commitment and enthusiasm Workload and resources Management systems Relevant experience Ability to innovate References	20–30
Project organization	Organization of project team Authority levels of team members Logistics related to site, client and other consultants Planning and programming expertise	15–25
Key project personnel	Qualifications and experience relevant to project Understanding of project brief Flair, commitment and enthusiasm Compatibility with client and other team members Communication skills References	30–40
Project execution	Programme, method and approach Management and control procedures Resources to be applied to the project Environmental, health and safety matters	20–30

The final quality/price assessment (Table 3.2) is achieved by multiplying the quality and price scores by the respective weightings set by the quality/price ratio and adding them together to give a total score out of 100 (e.g. if the quality/price ratio is set at 65/35, the quality score is 69 and the price score is 100, the total score is 69 × 65% + 100 × 35% = 79.9. The highest-scoring consultant should be awarded the contract.

The Construction Industry Council (a body representing all the professions) has developed a tool called *Selecting the Team* to help clients create a team able to work together successfully. It offers practical advice on how to put together a selection panel, develop a questionnaire, set the criteria for a shortlist and then evaluate the short-listed candidates. This publication complements two other CIC-partnering publications – *A Guide to Partnering Workshops* and *A Guide to Project Team Partnering*.

Basis of payment to consultants

There are three principal ways of paying for professional services (sometimes used in combination):

1. Time charge
2. Lump sum
3. *Ad valorem* – according to value.

Table 3.2 Example of completed tender assessment sheet.

Tender assessment sheet

Project: Halls of residence, University of Metropolis
Tenderer: bmg (Architectural services)
Assessor: Keith Potts
Project quality weighting: 65%
Project price weighting: 35%
Quality threshold: 65 (to be compared with total quality mark)

Quality threshold	Project weighting (A)	Marks awarded (B out of 100)	Weighted marks (A x B)
Practice or company	25%	64	16
Project organization	15%	80	12
Key project personnel	40%	65	26
Project execution	20%	75	15
	100%	Total quality score	69
Price criteria			
Tender price	260 (lowest compliant bid)		
Price score	100%	100	100

Overall assessment
Quality weighting x quality score = 65% x 69 = 44.9
Price weighting x price score   = 35% x 100 = 35.0
Overall score 79.9

Signed:
Date:

The fee structure to be adopted will depend on the degree of certainty in the scope and content of the services required. When the scope and content of the services are uncertain, for example, during the appraisal of options, then reimbursement on a time basis is appropriate. However, time charges provide no surety of the eventual fee cost. They tend to be an expensive way of paying for longer-term services and are more appropriate for shorter-term commissions.

Lump-sum charges should only be used where the scope of all the services is defined precisely and there is little risk of significant variations in the scope of the works. A combination of lump-sum charges for the more certain elements of the work, and time charges for those less certain, may offer best VFM.

Ad valorem fee structures reimburse consultants in proportion (generally as a percentage) to the cost of the project.

From the client's viewpoint they may appear at times to provide an incentive for consultants to design expensive projects rather than those offering VFM. It may therefore be appropriate to introduce some form of abatement or capping mechanism to the fee structure, in order to underline the necessity of striving to contain certain costs while maintaining quality. However, great care should be taken when developing such a model in order not to penalize those who are not responsible for changes or who have carried out abortive work or had to provide additional services as a result of the changes caused by other parties.

If the consultants' fees are calculated based on the final construction cost during times of high inflation then there may be an overpayment, known as an *uncovenanted gain*, as the majority of consultants' work is usually carried out during the early stages of the project.

Panel 3.1 Case study: Scottish Parliament building

On the Scottish Parliament building each consultant's fee remuneration was wholly or mainly a percentage value of the approved construction cost of the project.

In the 2004 Audit Report on the Scottish Parliament building, the Auditor General for Scotland identified that before they appointed consultants the client's project management could have explored more carefully the alternative fee arrangements with its consultants including final incentives linked to delivering value for money. 'Percentage fees do not align the objectives of the client with the commercial objectives of the consulting firms because the more a project costs the more each consultant is paid.'

In the event, a fee cap was agreed with cost consultants and the service engineer 12 months before completion. The final fees with the architect had not been finalized three months before completion whilst the final payment to the construction manager was uncertain due to the qualified nature of the agreed cap with them.

Source: Audit General for Scotland, 2004.

3.3 Selecting contractors by value

The principal aim of the selection process is to select a contractor who offers VFM. This will nearly always involve a process of competitive tendering. 'Value for money for a particular project means optimizing the balance between best performance or quality of service and lowest price' (OGC, *Successful Delivery Toolkit*, 2005).

A very useful approach to tender evaluation is to use a quality/price mechanism which is based on a numerical scoring and weighting system. The advantages of this approach include the following:

- It formalizes what can be a very subjective process;
- It requires forethought as to what are the most important criteria for selection;
- It can be transparent;
- It can be audited.

Traditionally, there have been three types of appointment procedure:

1. *Open tendering*: all interested parties can submit tenders in response to an advertisement in a local or national newspaper or the *Official Journal to the European Union* (OJEU) notice. Normally a small deposit is required which is refundable on submission of a bona fide tender.

    The open-tendering approach is an inefficient use of the industry's resource. It can lead to an excessive number of tenders with the lowest tender being a highly risky choice; it is not recommended.
2. *Selective or restrictive tendering*: allows the number of organizations to be restricted by using a selection process in advance of tender invitation. Under this approach client bodies who are regular users of the industry can keep standing lists, which should be reviewed annually, for different values and types of work. For major one-off projects, contractors can be selected for tender using a pre-qualification questionnaire.

3. *Negotiated tendering*: available in two forms:

    a. Competitive – enables clients to negotiate the terms with selected bidders and may include a formal tender stage prior to negotiation.
    b. Without a call for competition – used in only the most exceptional circumstances for example emergency storm-damage repair.

The selection process

The selection process produces a short list of the most suitable organizations from those that expressed an interest in carrying out the project. The selection process must be objective, fair, accountable and transparent with the criteria for selection established before inviting expressions of interest.

The selection process consists of the following steps:

*   Identify the selection criteria;
*   Establish the weighting;
*   Identify the minimum thresholds for selection criteria, where appropriate;
*   Construct the selection mechanism;
*   Invite expressions of interest;
*   Draw up a long list;
*   Draw up the short list.

The selection criteria for contractors should be based on attributes that fall under the following headings:

The contractor's personal and financial standing

Whether the contractor is financially stable and/or has the backing of a large group. Such assessment normally includes examination of accounts, annual reports (if a public company) or a confidential report from the company's bank.

Under this heading a company would not be suitable for selection on the following grounds: bankruptcy, failure to pay taxes, serious misrepresentation, grave misconduct or conviction of a criminal offence.

Technical and organizational ability

Whether the contractor has sufficient experience in the particular type and magnitude of works and has a satisfactory performance reputation in such areas as available resources; design expertise; experience of partnering, supply-chain management in force with subcontractors and suppliers; policy on risk management, for example, terms of subcontracts, history of rejected claims, skill/qualifications profile of workforce (own and key subcontractors), quality management; health and safety record.

Considerable care and effort will be required to set appropriate selection criteria for individual projects in order to make sure that only suitable contactors are selected.

Table 3.3 shows an example of a selection mechanism and demonstrates how the selection criteria and the weightings applied to them are used to evaluate each organization.

Table 3.3 Example of selection mechanism for contractors.

Project title: Baywatch Holiday Village
Assessors:
Overall quality threshold: 50
Mightybuild construction plc

Financial standing

Selection criteria	Quality threshold (QT)	QT reached?
Bankruptcy, convictions, misconduct etc.	Clean record	Yes
Profit and loss for last three years		Yes
Public liability insurance		Yes
Professional indemnity insurance (where applicable)		Yes

Technical and organizational

Selection criteria	Criteria weighting (a)	Score awarded (b)	Weighted score (a x b)
Relevant technical experience	20	80	16.00
Resources relevant to the project	10	70	7.00
Relevant design experience	15	60	9.00
Past performance on teamwork/partnering	15	30	4.50
Past performance on risk/value management	10	50	5.00
Rejected claims history	10	10	1.00
Health and safety record	10	70	7.00
Quality assurance	10	60	6.00
Total	100		55.50

The award process

The award process looks *forward* at the proposals for the specific contract whereas the selection process looks *back* at the status and previous performance of the bidders. The award criteria must be identified before inviting tenders and must be confirmed in the *instructions to tenderers.*

The award process is separate and distinct from the selection process. The award must be made on a *VFM basis* (value for money, that is, the most economically advantageous to the client) and not on lowest cost alone.

The award process comprises the following activities:

- Confirm candidates;
- Establish award criteria, weightings for award criteria, quality/price ratio, award mechanism, price scoring;
- Evaluate quality element;
- Prepare instructions to tenderers;
- Invite tenders;
- Evaluate *price* element;
- Balance quality and price;
- Notify award and debriefing.

Table 3.3 shows that Mightybuild Construction passed the initial financial-standing and technical/organizational tests with a score 55.50 (above the minimum threshold of 50) – they can therefore be considered for selection. However, it should be noted that although they have a good technical and design record, their performance on partnering and claims is not good. In fact, Mightybuild are currently involved in adjudication/litigation on 25% of their existing projects; potential problems may lie ahead for the client if Mightybuild are selected!

Evaluating quality

The quality evaluation of each short-listed firm will normally be based on a pre-tender interview, which is often preceded by the completion of a pre-interview questionnaire. The quality scores should be established before the price bids are opened. Mightybuild usually score well at such interviews, indeed their competitors have observed that Mightybuild have a policy of sending the 'A' team to interviews and the 'Z' team to carry out the project!

Topics likely to be considered within *Quality* include the following:

- *Team-working arrangements*: partnering with client and subcontractors;
- *Aesthetic and functional characteristics*: design, operating costs, ease of use, adaptability, innovation, maintainability;
- *Proposals for managing the contract*: planning, programming, management, milestones for achieving objectives, risk identification and proposals for management, communication, quality plan;
- *Project team organization*: qualifications and experience of team members, senior managers, partners, quality of senior personnel, resources;
- *Technical merit*: proposed methods, approach to CDM Regulations, health and safety management, the design and construction stages, quality of documentation, standards of materials, checks and inspections;
- *Services provided from external sources*: details of joint-venture proposal, arrangements for subcontracting, training amongst workforce in the supply chain.

Table 3.4 indicates typical quality/price ratios for different types of project.

Evaluating the price

There are various methods of evaluating a contractor's tender price; one method requires the following approach:

- The *mean* price of the acceptable tenders is given 50 points;
- One point is deducted from the score of each tenderer for each percentage point above the mean;
- One point is added to the score of each tenderer for each percentage point below the mean.

Unjustifiably low tenders should be rejected and not included in the assessment.

Table 3.4 Typical quality/price ratios for different types of projects.

Type of project	Indicative quality/price ratio
Innovative projects	20/80 to 40/60
Complex projects	15/85 to 35/65
Straightforward projects	10/90 to 25/75
Repeat projects	5/95 to 10/90

So in the example shown in Table 3.5 the mean price is £7,717,000

Alpha   50.00 + 6.05 = 56.05
Mightybuild   50.00 − 8.20 = 41.80
Zed construct   50.00 + 2.16 = 52.16

Comments: After multiplying the *quality* scores by 60% and the *price* scores by 40% the award mechanism shows that the contract should be awarded to Mightybuild (Table 3.5).

Table 3.5 Award mechanism.

Project title: Baywatch Holiday Village
Quality/price weighting: 60:40
Assessors:

Quality scores

Quality criteria	Criteria weight (%)	Alpha construction		Mightybuild construction		Zed construct	
		Score	Wtd score	Score	Wtd score	Score	Wtd score
Proposals for and understanding of project	30	60	18.00	80	24.00	70	21.00
Experience and resources of proposed project team	20	55	11.00	90	18.00	50	10.00
Project management/ team working skills	10	70	7.00	60	6.00	50	5.00
Risk management skills and experience	10	55	5.50	70	7.00	60	6.00
Aesthetic character of proposals	15	70	10.50	70	10.50	80	12.00
Maintainability	15	50	7.50	60	9.00	70	10.50
Total quality			59.50		74.50		64.50

Price scores

	Alpha construction	Mightybuild construction	Zed construct
Tender price	£7,250,000	£8,350,000	£7,550,000
Total Price (mean £7,717,000)	56.05	41.80	52.16

Overall scores

	Alpha construction	Mightybuild construction	Zed construct
Quality weighting x quality score	60% x 59.50 = 35.70	60% x 74.50 = 44.70	60% x 64.50 = 38.70
Price weighting x price score	40% x 56.05 = 22.42	40% x 41.80 = 16.72	40% x 52.16 = 20.86
Overall score	58.12	61.42	59.56
Order of tenders	3	1	2

However, in compiling the table, the author identified two major areas of difficulty in developing the quality part of the model, particularly:

1. identifying the key criteria and the relevant weighting;
2. identifying a realistic score against the criteria for each contractor.

3.4   Construction Industry Research and Information Association (CIRIA) Guide – *Selecting Contractors by Value*

In 1998 the CIRIA produced the definitive guide *Selecting Contractors by Value* (Jackson-Robbins, 1998). The guide provides an overview of the key processes in selecting contractors by value.

The guide identified the following key issues to be considered in the selection of contractors:

1. Defining value – what represents value to the client?
2. Identifying opportunities for contractors to add value – where can contractors add value to the project?
3. Developing the procurement strategy to secure value – when and how can contractors be involved?
4. Defining the selection criteria – how can potential contractors be judged?
5. Obtaining information – how can a full picture be obtained of potential contractors?
6. Making the choice – how can a balanced and accountable selection be made?

The CIRIA guide includes a toolbox comprising 12 matrices which are intended to help clients and their advisers compare the ability of potential contractors to add value to a project.

Matrix 1: technical knowledge and skill;
Matrices 2–8: skill and commitment in managing time, cost, quality, risk, health and safety, environmental issues;
Matrix 9: effectiveness of contractor's internal organization;
Matrix 10: contractor's attitude and culture;
Matrix 11: quality of human resources proposed (see Table 3.6);
Matrix 12: quality of supply-chain management.

Each matrix contains up to six indicators against which there are definitions allowing the scorer to judge the contractor to be *poor, adequate, good* or *excellent.*

The total scores can be calculated against each matrix to arrive at a total weighted score similar to Table 3.2. The CIRIA toolbox removes some of the subjectivity involved in the process by giving definitions against each attribute, thus making the process more transparent and auditable.

Significantly, the author of the CIRIA guide admits that the final scores of even the most sophisticated matrix should only be used as an aid to making a judgement. The guide advises that the selection should critically assess the numerical results and if necessary test them, for example, by carrying out a sensitivity analysis. Would you recommend Mightybuild?

Table 3.6 Matrix 11: human resources – indicator 4.

Company employment policy	Regular use of agency staff	Staff regularly employed, with reasonable lengths of service	Company training policy, including staff appraisals/development programme	'Investors in people' award, loyal and enthusiastic employees
Overall assessment	Poor (−1)	Adequate (0)	Good (1)	Excellent (2)

Source: CIRIA, 1998.

Panel 3.2 Case study: Stoke-on-Trent Cultural Quarter

The appointment of an appropriate management contractor was a key decision for this project. The Council made use of a methodology, Most Economically Advantageous Tender (MEAT) that was starting to be used in the public sector.

The MEAT methodology scores each bid against a range of qualitative and quantitative factors. The Council officers decided that quality would account for 60% of evaluation, which would be assessed at interview and included previous management contracting experience and experience of working on theatres and concert halls. Cost would count for 40%, the calculation being based by the extent, which the bids were above or below the average cost of all the bids.

The Council invited five potential contractors to attend interviews and submit tenders – one contractor was deleted.

	Norwest Holst	Tilbury Douglas	Bovis	Sunley Turriff
Interview	344	380	470	362
Tender	300	225	113	163
Total	644	605	583	525
Fee	£1,264,188	£1,398,973	£1,602,924	£1,510,938

Norwest Holst secured the highest number of points: 644. However, they were ranked last in relation to quality assessment. The procurement route was non-traditional with a high degree of risk, therefore, the contractor's experience of management contracting was essential. Bovis had the most experience of both management contracting and refurbishment of theatres, Norwest Holst had the least relevant experience – yet were selected.

The project cost the Council £15 million more than the original budget of £22 million yielding a total spend of £37 million.

For a fuller description on this case study see Chapter 11 (Panel 11.2).

Source: Audit Commission: District Auditor's Report on Cultural Quarter Stoke-on-Trent City Council, 22 January 2004.

The Stoke Cultural Quarter case study indicates the care which must be taken in the initial selection of contractors.

In November 2005, ICC Credit one of the UK's leading credit-reference agencies reported that Sunley Turriff Construction Limited, based in Manchester, had gone into liquidation. The company was established in 1993. Over the 12-year period, the company went through a number of ups and downs. In 1998 the company achieved its highest turnover of £95 million. However, its profit record failed to perform as well. Over the last seven years the company was not profitable and in 1995, its profit slumped to a negative figure of £4.5 million and remained in the red for the rest of its existence (www-icc-credit.co.uk [accessed January 2007]).

Using the selection methodology described in the Baywatch Holiday Village example, Sunley Turriff would not have passed the financial standing section within the quality threshold and so would not have been included in the short list.

3.5   Two-stage tendering

Two-stage tendering is a procedure typically used to secure the early appointment of a contractor under a lump-sum contract. The main benefits of this approach are early completion and the potential to include the contractor's buildability expertise within the design. In addition, this approach allows the client to have a greater involvement in the pre-selection and appointment of subcontractors.

The two-stage process involves the following:

1. *Pre-qualification*: main contractor tenderers;
2. *Compile first-stage tender based on*: programme – method statement, pre-construction fee, preliminaries – overheads and profit – initial pricing of packages;
3. *Identification of preferred contractor*: pre-contract services agreement;
4. *Pre-qualification*: tender of subcontract packages;
5. *Subcontractor*: selection by employer and contractor (allows for novation of client-appointed specialists);
6. *Compile second-stage tender*: first-stage contractor; agreement of subcontract terms; risk allowances;
7. Agreement of second-stage, lump-sum tender;
8. *Award of main contract*: commencement of works and subcontractor appointment.

The second stage, which is typically managed as negotiation between employer and the preferred contractor relies on competition between second-tier contractors for work packages.

The two-stage tendering, should in theory, be well suited to design and build projects. However, while this approach should give the client additional control over the design development and the transfer of risk to the contractor, it normally comes at a premium price. For a fuller commentary on two-stage tendering see Simon Rawlinson's article in the *Building* magazine, 12 May 2006.

An economic market forecast by Davis Langdon in the *Building* magazine reported that in London 'Contractors are unwilling to take unattractive procurement routes. It is difficult to attract interest in single stage design-and-build projects with contractors more interested in two-stage design and build on traditional routes. Furthermore, two stage tenders have increasingly become contractors' preferred approach for complex projects and some projects cannot be let by any other means' (Fordham, P., 2007).

3.6   FIDIC tendering procedures

The FIDIC document *Tendering Procedure*, 2nd edition (FIDIC 1994) presents a systematic approach for tendering and awarding of contracts for international construction projects. Experience has shown that pre-qualification is desirable since it enables the employer/engineer to establish the competence of companies subsequently invited to tender.

The *Tendering Procedure* shows the following flowchart of activities:

1.0 Establishment of project strategy;
2.1 Preparation of pre-qualification documents;
2.2 Invitation to pre-qualify;
2.3 Issue and submission of pre-qualification documents;
2.4 Analysis of pre-qualification applications;
2.5 Selection of tenders;
2.6 Notification of applicants;

3.1 Preparation of tender documents;
3.2 Issue of tender documents;
3.3 Visit to site by tenderers;
3.4 Tenderers queries;
3.5 Addenda to tender documents;
3.6 Submission and receipt of tenders;
4.1 Opening of tenders;
5.1 Review of tenders;
5.2 Tenders containing deviations;
5.3 Adjudication of tenders;
5.4 Rejection of all tenders;
6.0 Award of contract;
6.1 Issue letter of acceptance;
6.2 Performance security;
6.3 Preparation of contract agreement;
6.4 Notification of unsuccessful tenderers.

Under section 5.3 Adjudication of tenders, the tendering procedures identify that the evaluation of tenders are generally considered to have three components:

Technical evaluation

- conformity with specification and drawings;
- comparison of any proposed alternatives (if allowable) with the requirements of the tender documents;
- design aspects for which the contractor is responsible;
- methods of construction and temporary works;
- environmental considerations;
- quality assurance;
- programme.

Panel 3.3 FIDIC tendering procedure

The FIDIC document *Tendering Procedure* gives guidelines on the analysis of pre-qualification applications – the evaluations should determine, for each company or joint venture:

- Structure and organization;
- Experience in both type of work and the country or region in which it is to be undertaken;
- Available resources, in terms of management capability, technical staff, construction and fabrication facilities, maintenance and training facilities, or other relevant factors;
- Quality assurance procedures and environmental policy;
- Extent to which any work would be likely to be subcontracted;
- Financial stability and resources necessary to execute the project;
- General suitability, taking into account any potential language difficulties;
- Litigation or arbitration history.

Source: FIDIC, 1994.

Financial evaluation

- capital cost;
- discounted cash flow and NPV;
- programme of payments;
- financing arrangements;
- currencies;
- securities;
- interest rates;
- down payments/retentions;
- daywork rates;
- contract price adjustment proposals.

General contractual and administrative evaluation

- conformity with instructions to tenderers;
- completeness of tenders;
- validity of tenders;
- exclusions and deviations – stated or implied;
- insurance;
- experience of key staff;
- shipping, customs, transport;
- working hours;
- labour build-up, run-down and source.

It is noted that the FIDIC contract highlights the importance of assessing tenders on a net present value (NPV) basis. The technique is also recommended by the World Bank yet it is rarely encountered in standard UK contracts. This powerful tool enables an equitable comparison to be made of all the tenderers' forecasted project cash flows on a present-day cost basis. The challenge for the client's commercial advisers is devising a tendering system that allows an NPV comparison to be made.

Panel 3.4 Case study: tender assessment – international mega project based on FIDIC

The assessment of tenders was carried out in two stages. The first stage involved a coarse assessment of all tenders to identify the tenders to be considered in depth. The second stage involved a detailed appraisal and examination of the most favoured tenders together with negotiation with tenderers, as appropriate, leading to final recommendation.

The tender assessments were carried out by groups reporting to the Chairman and Executive. The activities involved included the following:

Finance department

Examine financial package; check front loading; bring all tenders back to Net Present Value utilizing the Interim Payment Schedule included in the tender; arithmetical check of BofQs; check alternative offers; identify most favourable in financial terms.

Contracts/legal department

Examine the covering letter and all qualifying statements; indicate cost and time implications of qualifications.

Programming department

Check that the tenderers programme adheres to schedules of milestones and critical dates identified in the contract documentation.

Consultants/civil engineering department

Examine alternative design offers and recommend any which warrants further consideration and advise on any planning implication; examine method statement to highlight any anomalies and areas of non-acceptance; prepare detailed cost comparison to enable cost centres and activity bills to be compared with employer's estimate.

Utilities and civil planning

Evaluate the implications of the tenderers submission on traffic, land access and utility diversions.

   At the second-stage assessment, a detailed study would be undertaken and questions of clarification would be developed. Following the issue of the questions to the tenderers and receipt of the answers, all outstanding matters would be clarified. The confirmation wrap-up letters would be incorporated into the contract.

3.7   Conclusion

We have now examined the procedures for selection of both consultants and contractors in the UK and overseas. In the UK, as part of the VFM process demanded by the government, there has been a shift by public sector clients to selection based not only on lowest price but also embracing *quality* issues.

   Over the past decade many significant guides have been published by the Construction Industry Council and CIRIA in order to assist clients and their consultants in developing best practice. However, the Baywatch hypothetical case study and the Stoke-on-Trent Cultural Quarter case study demonstrate that even when *quality* issues are incorporated into the selection process of the project, major problems can still arise.

3.8   Questions

1.   The University of Metropolis wishes to appoint the client's project manager with full responsibility for controlling and managing the time, cost and quality on the new £25 million Learning Centre and Teaching facility. The Executive of the university understands that the construction industry is now committed to providing best value for its clients and is keen that this philosophy is adopted by all its consultants.

   Describe the pre-qualification selection process and identify the criteria for the short list to tender.

2.   Access the OJEC website (www.ojec.com) and identify the following:

   a.   What does OJEU mean?
   b.   Which projects are covered by its regulations?
   c.   What are the public procurement thresholds?
   d.   What criteria are used to evaluate tenders?
   e.   What process is followed?

3.   In June 1998 the Foreign and Commonwealth Office (FCO) signed a contract for the construction, operation and financing of a new embassy in Berlin with a German supplier called Arteos which had been formed by a consortium comprising Bilfinger + Berger, one of the three biggest German construction companies, and Johnson Controls, a large American-based facilities management provider. The construction was completed in 2000. Review the techniques used in the selection of Arteos on this PFI project (www.nao.org.uk/publications/nao_reports/9900585.pdf [accessed 4 February 2007]).

Bibliography

Auditor General for Scotland (2004) *Management of the Holyrood Building Project*, Audit Scotland

Construction Industry Board (1996) *Selecting Consultants for the Team: Balancing Quality and Price*, CIB

Construction Industry Council (1994) *Procurement of Professional Services. Guidelines for the Value Assessment of Competitive Tenders*, CIC

Construction Industry Council (2005) *Selecting the Team*, CIC

FIDIC (1994) *Tendering Procedure*, 2nd edition, FIDIC, Switzerland

Fordham, P. (2007) 'Economics: market forecast', *Building magazine*, 2 February, pp. 64–65

Holt, G.D. (1998) 'Which contractor selection methodology', *International Journal of Project Management*, vol. 16, no. 3, pp. 153–164

Holt, G.D., Olomolaiye, P.O. and Harris, F.C. (1995) 'A review of contractor selection practice in the UK construction industry', *Building and Environment*, vol. 30, no. 4, pp. 553–561

Jackson-Robbins, A. (1998) *Selecting Contractors by Value*, Construction industry Research and Information Association

National Audit Office (NAO) (2000) 'Report by the Comptroller and Auditor General', The Foreign and commonwealth Office, The New British Embassy in Berlin, HC585, Session 1999–2000, The Stationery Office, London

Rawlinson, S. (2006) 'Two-stage tendering', *Building* magazine, 12 May, pp. 62–65

4 Pre-contract cost management

4.1 Introduction

Cost management is the process, which is necessary to ensure that the planned development of a design and procurement of a project is such that the price for its construction provides value for money (VFM) and is within the limits anticipated by the client.

Construction is a major capital expenditure, which clients do not commence until they are certain that there is a benefit. This benefit may be for society in the case of public projects, with justification based on a cost–benefit analysis, or purely based on financial considerations in the case of private projects.

Most clients are working within tight pre-defined budgets, which are often part of a larger overall scheme. If the budget is exceeded or the quality not met the scheme could fail. Pre-contract estimating sets the original budget – forecasting the likely expenditure to the client. This budget should be used positively to ensure that the design stays within the scope of the original scheme.

When developing an estimate the following factors need to be considered:

- Land acquisition including legal fees;
- Client's own organization costs allocated to the project (this obviously varies but can be as much as 10% of the overall project budget);
- Site investigation (frequently underrated and under-budgeted resulting in unnecessary extra costs and time – could be as much as 1% of budget);
- Enabling works, decontamination;
- Insurances (many major clients prefer to insure against the risks and take out a project insurance policy covering both themselves and the contractor – may be up to 1% of the budget);
- Consultants' fees including design (on large transportation and infrastructure projects this can be as much as 15–20% of the budget);
- Construction costs (typically account for between 70% and 80% of the project sum) (excluding land);
- Value added tax (VAT) (currently charged at 17.5%);
- Contingency and risks (covers for the unknown and may be between 20% and 25%) or if project of long duration the contingency factor could be double or triple these items;
- Financing and legal costs (financing costs can be substantial depending on financing method chosen and typical bank rate – could be anything between 7% and 20%; lawyers are expensive – anything up to £500 per hour and more).

Cost control has to be exercised before any commitment is made. To do otherwise sees cost control become a procedure of cost monitoring only. Pre-contract financial control therefore should be a proper mix between design-cost control and cost monitoring but with the emphasis on positive cost control rather than passive monitoring. An essential tool for financial control is the cost plan.

However, in practice, the difficulties of estimate production is exacerbated when the project involves major uncertainties; perhaps because no similar work has been tackled before, or because the scope of the work is poorly defined (Swinnerton, 1995). The Scottish Parliament Building is a prime example being a unique building with a poorly defined brief – initial budget £40 million and the final cost reported to be £431 million.

4.2   Cost estimating on engineering, manufacturing and process industries

Estimates of the cost and time are prepared and revised at many stages throughout the project cycle. These are all predictions and should not be considered 100% accurate. The degree of realism and confidence achieved will depend on the level of definition of the work and the extent of the risk and uncertainty. Consequently, as the design develops, the accuracy of the estimate should improve.

The Joint Development Board's publication, *Industrial Engineering Projects,* demonstrates this principle of estimating accuracy clearly in connection with capital projects in Engineering, Manufacturing and Process Industries (JDB, 1997; see Fig. 4.1). The JDB is sponsored by the Royal Institution of Chartered Surveyors and the Association of Cost Engineers, and is charged with raising the profile of project and commercial management in the engineering industry.

The JDB document identifies four main types of estimate, which could be produced by the owner or contractor, each with a different level of accuracy and each used at different stages throughout the project cycle.

Order of magnitude estimate

This order of magnitude or *ballpark* is produced for the rapid evaluation of commercial possibilities and economic viability of a project. Since little detail will be normally available the estimate will usually be based on data from a similar previous project updated for time, location, changes in market conditions, current design requirements and relative capacity.

In the absence of data from a near-duplicate project, the estimator will rely on published or historical data from a number of existing projects, usually related to the overall size or capacity of the project or facility concerned, adjusted as necessary. An order of magnitude estimate will typically have an accuracy of −25% to +50% (JDB, 1997).

Typical examples of this type of estimating include the following:

- Cost per megawatt capacity of power stations;
- Cost per kilometre of highway;
- Cost per tonne of product output for process plants;
- Cost per car park space (multi-storey car park), pupil (school), beds (hospital) etc.

The key issue to consider when using this approach is comparing like with like: Are the standards the same in the previous projects? Does the price include infrastructure? Are professional fees and financing costs included etc.? Despite these concerns an order of magnitude estimate can be useful, particularly at the conceptual stages of projects when information is very limited and alternatives have to be ranked quickly (Norman, 1994).

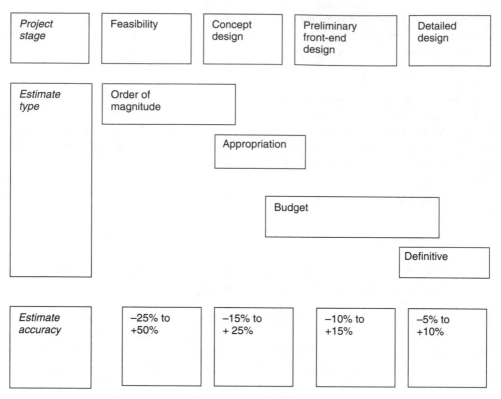

4.1 Estimating accuracy in industrial engineering projects (source: JDB, 1997).

This approach is probably realistic for all complex major projects, including civil engineering and building. It is based on the concept that the degree of accuracy of the estimate is only as good as the level of detail available. In practice, clients often demand certainty of outcome from inception requiring the design team to successfully manage the development of the design within the initial budget.

Appropriation estimate

In the engineering and process industries, the appropriation estimate is sometimes referred to as the Class III estimate as it uses information developed to a level of definition described as Class III. At this stage the designers will have identified the major equipment and determined their required outputs. This will provide an opportunity to enable the estimator to make enquiries of potential suppliers regarding the availability and price of key components. The appropriation estimate will typically have an accuracy of −15% to +25% (JDB, 1997).

Budget estimate

In the engineering and process industries, the budget estimate is sometimes referred to as the Class II estimate and is produced once the conceptual design has been completed. The budget estimate will typically have an accuracy of −10% to +15% (JDB, 1997). In those same industries information available at this stage will allow for approximate quantities to be established and guide prices obtained from potential vendors.

Definitive estimate

In the engineering and process industries, the final estimate produced immediately following commitment to the major capital expenditure is the definitive or Class I estimate with an accuracy in the range of −5% to +10% (JDB, 1997).

This estimate will typically contain the level of detail used in the execution of projects and the preparation of bids. It will be used in the maintenance of close control over the cost of the work, or for allocation of resources into work packages.

The order of magnitude estimate takes a *top-down* approach, probably based on the final cost of previous projects. In contrast, the definitive estimate takes a *bottom-up* approach; the estimate is built up from specific project information and realistically may therefore contain omissions for risks and uncertainties.

4.3   Cost estimating on civil engineering projects

In carrying out cost management there should be a clearly defined route from feasibility stage through to the placement of a contract. There should be break points, or gateways, when the client can take the decision whether to proceed or not. This is in line with the recommendations by the Office of Government Commerce in their Gateway Review Process.

One of the benefits of cost management in the pre-contract stage, especially in multi-contract projects, is that it helps the project team to better establish the appropriate project contract strategy. That is, which work should be placed in which contract and possibly the form of contract which should be adopted for particular contracts. Cost management can also help identify possible programme restraints both in contract preparation and execution.

The preparation of the first estimate would be based on a variety of techniques, for example, historic data or approximate quantities. Major projects often have substantial elements that are unique and for which there is no relevant historic data. In these cases it is necessary to analyse the project in as many individual work sections as can be identified, if possible to prepare indicative quantities and consider the resources necessary to carry out the work. During this indicative stage it is wise to contact potential contractors and manufacturers especially with regard to order-of-cost estimates for specialist sections.

Other matters that have an effect on cost and need to be addressed at this time include location of project and access thereto, especially with regard to heavy and large loads, availability of labour and the possible need for residential hostels or other accommodation for workmen, off-site construction, temporary works. It will also be necessary to consider allowances for design development, allowances for consultants' fees and client's costs, land-acquisition costs and general contingencies.

When the client has accepted the first estimate and instructs that the project proceed to the next stage, then this becomes the first cost plan against which further design developments and changes are monitored.

During the process of design development the main duties of the quantity surveyor as part of the cost management team are as follows:

- to check and report the cost of design solutions as they are established or refined by the engineers;
- to prepare comparative estimates of various design solutions or alternatives and advise the engineer accordingly;
- as changes are introduced into the project, to estimate the cost effect of the change and to report;

- to prepare a pre-tender estimate based on a bill of quantities (BofQ) or priced activities;
- to prepare a financial appraisal.

The monthly issue of the updated cost plan is the vehicle whereby the cost management team is made aware of the current estimated cost of the project. In its simplest form a pre-contract cost plan will set out in tabular form each and every work section, the approved estimate for that section, the estimate for the previous and the current month for the section and a note of the changes that have taken place in the month. The total of all the sections provides the estimated cost of the project.

There should be a continuous dialogue between the designers and the quantity surveyor (QS); ideally both should work together in the same office during the critical stages of design development. Normally, there are so many changes within a month during design development that these are better listed as an appendix to the cost plan. One national client insists that a separate appendix to the cost plans lists all potential changes and these have to be approved by his project manager before changes can be included in the cost plan. In this way the cost plan represents committed cost only (Shrimpton, 1988).

The extent of detail in the preparation and updating of cost plans is such that it is best handled on a database for transfer to a spreadsheet.

The accepted estimate in the form of priced activities or BofQs becomes the basis for the first post-contract cost plan. This then acts as the client's design datum for cost management and reporting in the construction stage.

Highway works

An approximate estimate of the cost of constructing major highway works is usually required at an early stage in the project cycle in order to determine if the scheme is reasonable and will fit in with government-funding allowance. At this early stage the proposed project will be analysed in fundamental elements. The road construction will be estimated at £x per linear metre for three-lane, dual or single carriageway – this price will normally include drainage, lighting and signage.

The estimate will be based on an analysis of previous similar tenders using the highway consultant's own data or data from personal contacts, or following up leads in government White Papers or in the *New Civil Engineer*. Adjustments need to be made for inflation and market forces using the Department of Transport's *Road Construction Price Index*, which is published quarterly and shows trends in national prices.

Additional items to be considered include earthworks, statutory undertakers' equipment to be moved, townscaping and landscaping, telephones, closed-circuit TV etc. Bridges are kept separate and are again estimated based on the consultant's own cost data, taking into account the number of spans/type of construction/length/width of carriageway etc.

The Highways Agency (HA), acting as Executive Agency for the Department of Transport has radically changed its procurement policies following the recommendations made in Egan's report *Rethinking Construction*.

Since the mid-1990s the HA has undertaken most major projects using design and build (D&B) contracts with most risks transferred to achieve greater certainty of spend. The scope for contractor innovation has been limited because they have not been appointed until after the statutory planning stages that establish many constraints.

In addition, improved price certainty has been sought by transferring risks, without giving full recognition to a contractor's ability to assess and manage the risks. The HA now recognizes that this approach does not always support partnership working if commercial pressures come to the fore. Improved VFM can be achieved by allocating risks appropriately, and price certainty delivered by managing the risks in partnership, supported by incentives.

Panel 4.1 Highway scheme estimates

In 2006 *The Times* reported that highway schemes to ease congestion are being stalled and many could be scrapped. According to the article one-third of planned improvement schemes have been delayed up to five years.

> The Highways Agency said that the nature of the programme meant that some schemes could fall behind or progress quickly. But construction companies blame Treasury rules for budgeting, which force the Highways Agency to include a margin of error of up to 45 per cent.

> A Highways Agency spokesman said that the Government was committed to the programme of improvements and that all the roads would be built. The agency said that it was working with contractors to reduce costs. Where that was not possible, increases would be met from a special reserve.

(Jameson, A. and Nugent, H., 2006)

Note: This approach seems to be in line with the JDB recommendations on order of magnitude estimates at feasibility stage in the range −25% to +50% accuracy.

The earlier appointment of a contractor offers considerable scope for better value, but it is important to get the right timing. The earlier it is, the more scope there is for the contractor to contribute expertise and innovation, but the time period for construction should not be too long.

The use of project-partnering arrangements on the HA's major projects in recent years has been beneficial in achieving mutual objectives for the particular projects. However, the procurement of major projects on an individual scheme basis means that the partnerships and the invested knowledge and experience of team members can be lost to the client, if there is no continuity of work. The lack of continuity also makes it difficult for suppliers to plan their resources and does not encourage the training and development of the workforce. HA now recognizes that this could be resolved by applying long-term relationships to the delivery of major projects.

HA has identified that the following actions need to be taken on D&B projects:

- For publicly funded major projects the HA will normally use a new form of D&B contract, known as early D&B, where the contractor is involved much earlier in the planning process. The contractor will be appointed as soon as possible after identification of the preferred route and well before the statutory stages which normally involve a public inquiry.
- The contractor selection process will be based largely on quality, with the HA seeking to identify a supplier that has all of the right skills and who is considered most capable of working in partnership, to identify the optimal solution and to deliver it as efficiently and safely as possible.
- Suppliers will need to demonstrate good supply-chain management practices as set out in this document. In particular, the relationship between the contractor and their designer will be very important and the HA will require designers to be adequately incentivized to deliver optimal solutions.

- The burden of tendering will be kept to a minimum by avoiding the need, as far as possible, for detailed design work during the tender stage.
- Pricing will be based on key cost components and a process to establish a target cost when the design is finalized. Target costs will be incentivized in a way that encourages continuous improvement throughout the development of the project.
- Risk schedules will be developed with tenderers as part of the quality-assessment process and also to identify a fair allocation of risks to the parties best able to manage them.
- The HA will develop ways of entering into longer-term relationships with contractors on the delivery of major projects to achieve the benefits which are being achieved on new main-tenance contracts and framework arrangements. Options that will be examined will include frameworks and the packaging of projects into long-term programmes (www.highways. gov.uk/business/931.aspx [accessed 20 January 2006]).

This change of approach will have a significant impact on the cost-control process throughout the pre-tender period. Critically the design team, including the contractor and supply chain, will be working to establish innovative designs within the targets set for the key components.

4.4   Cost estimating on building projects

During the late 1950s the technique of elemental cost planning on buildings was established. This technique enabled the client to obtain a more reliable pre-tender estimate and gave the design team a template in order to control the cost during the design development stages. The technique was embraced by the Hertfordshire County Council and used successfully on the CLASP modular school building projects in the 1960s.

The technique is now well established in the building sector and has been further developed by the Building Cost Information Service of the RICS (BCIS) to include a national database of elemental cost analyses, which can be accessed using online computer techniques. Such information can be used to aid the pre-contract estimating process in the building sector as well as helping to ensure VFM by aiding the designer to ensure the most appropriate distribution of costs within the project.

Cost management is the total process, which ensures that the contract sum is within the client's approved budget or cost limit. It is the process of helping the design team design to a cost rather than the QS costing a design.

The basis of the design cost control using the cost-planning technique is the analysis of existing projects into functional elements in order to provide a means of comparison between projects planned with data from existing projects. A building element is defined as part of a building performing a function regardless of its specification. Elemental analysis allows the comparison of the costs of the same element to be compared between two or more buildings.

As the cost element under consideration is performing the same function, an objective assessment can be made as to why there may be differences in costs between the same elements in different buildings. There are four main reasons why differences in costs occur:

1.   Differences in time (inflation)
2.   Quantitative differences
3.   Qualitative differences
4.   Differences in location.

On a major project it is necessary to consider individual buildings or parts of buildings. A major shopping centre may be split into common basement, finished malls, unfinished shells, hotel and car parking. The parts of the whole may be physically linked and difficult to separate, but separation will ease estimating and control. The costs of the identifiable parts can then be compared against other schemes. For example, a composite rate per square metre is meaningless when you mix the cost of finished atrium malls with unfinished shells.

It is not only important to separate out parts of the building that serve different functions but it is equally important to separate for phasing. Many major projects have to be built around existing structures, which increase the cost because of temporary works as well as inflation.

The client's and project's status with regard to VAT will also need to be established. In the UK VAT is currently payable on building work other than constructing new dwellings and certain buildings used solely on both residential and non-business charitable purposes and also on all consultants and professional fees. The current VAT rate is 17.5%.

For further information on the application of VAT to construction works in the UK visit HM Revenue & Customs website (www.hmrc.gov.uk).

It is customary to exclude this amount from estimates and tenders, a practice that is well understood in the construction industry. However, this must be pointed out to any client who otherwise may think that the estimate is their total liability (Ferry and Brandon, 1999).

Design stages

References to design stages are to the RIBA Plan of Work (taken from the RIBA *Handbook of Architectural Practice and Management*) and refer to the main stages through which a project design typically passes. Table 4.1 shows the control tasks and deliverable reports required from the QS within the RIBA Plan of Work stages.

It is also recommended that value management and risk management are also carried out throughout the design process. These might affect both the client's requirements and the chosen design solution, and changes would, therefore, affect the budget and cost plan.

If at any time during the design process it becomes apparent that the agreed budget is likely to exceed without a change to the brief, the client should be informed and instructions requested. Likewise, if it becomes apparent that the whole of the agreed budget will not be required, the client should be informed.

Budget-estimating techniques

On projects where non-traditional procurement routes are used, the responsibility for developing the cost plan may change but the stages suggested here remain appropriate. For example, on D&B schemes, the client's QS will be responsible for the cost plan at feasibility and outline

Table 4.1 Control tasks and deliverable report.

Design stages	Quantity surveyor
Stage B: Feasibility	Prepare feasibility studies and determine the budget
Stage C: Outline proposals	Consider with client and design team alternative strategies and prepare outline cost plan
Stage D: Scheme design	Carry out cost checks and finalize cost plan
Stage E: Detail design	Carry out cost checks
Stage F: Production information	Carry out cost checks
Stage H: Tender action	Prepare reconciliation statement

proposal stage and the D&B contractor's QS will be responsible for developing the cost plan with the contractor's design team to produce the tender. However, the principles of budget, cost plan, cost checks and reconciliation should be adhered to whenever possible.

There are four main ways to estimate the cost of a building during the design stage, which are dependent on the quantity and quality of the information available at the time the estimate is required:

1.   Function or performance related
2.   Size related
3.   Elemental cost analyses
4.   Unit rates.

As a general principle, estimates prepared on minimal information and those taking a very short time to prepare will be less accurate than those based on the use of substantial information requiring more time to prepare.

The term *cost modelling* is often used to describe the function of cost estimating. Any form of cost prediction can be described as a cost model, whether it is based on functional, performance-related, elemental cost analysis or detailed rates calculated by contractors when pricing tenders. However, cost modelling generally implies the use of computer aids in order to allow iterations to be rapidly performed to help select the most appropriate solution to achieve VFM.

Functional or performance-related estimating

A function or performance-related estimate typically requires one quantity and one rate and is related to the client's basic requirement. Typical examples include the following:

*   A 1,000-bed hotel
*   A 2,000-pupil school
*   A 1,500-bed hospital.

For example, an hotelier will know that a hotel will cost £75,000 per bed to build and will earn £75 per bed per night. They can use this information to calculate the relative efficiencies of two proposed hotel options of completely different sizes.

An estimate based on this technique is very simplistic, crude but of course quick. It does not take into account plan shape, number of floors, ground conditions etc. It is considered extremely risky to use this technique except at the very earliest stages of inception. Often statistical techniques are employed in an attempt to improve the accuracy and reliability of the estimate.

Using this approach it would have been possible to identify a *ballpark* figure for the anticipated cost of the new Wembley Stadium at an early stage in the project cycle based on the cost per seat (refer Table 4.2).

An independent VFM study by surveyors Cyril Sweet in April 2002 described the new Wembley Stadium and identified some of the key features of this design and construction contract with multiplex as follows:

Value for money both in terms of the market price for the scheme as designed, and in terms of comparison with stadia of similar standing.

This does not signify that it is the cheapest price, but that it falls within the broad cost parameters for a project of this type and scale.

Table 4.2 Major stadiums cost-per-seat comparisons.

Details	Wembley	Stade de France	Stadium Australia	Munich (new stadium)	Sapporo Dome (Japan)	Washington State
Location	London, UK	Saint Denis, France	Sydney, Australia	Germany	Sapporo, Japan	Washington, USA
Capacity (seats)	90,000	80,000	80,000	66,000	42,122	72,000
Accommodation area (m^2)	73,000	70,000	100,000	—	53,800	—
Area per seat (m^2)	1.92	0.88	1.25	Not known	1.28	Not known
Total cost (£ million)	£352.6	£266.6	£278.9	£248.2	£246.0	£359.6
Cost per seat	£3,918	£3,332	£3,486	£3,761	£5,839	£4,995

Source: Report on Wembley Value for Money produced by Cyril Sweet Ltd on 10 April 2002 (www.sportengland.org/ new_wembley.pdf [accessed 23 January 2006]).

We consider that the level of the Multiplex price reflects the specific aspects of the proposed Wembley contract. Onerous conditions of contract, developed to provide greater security of outturn price for WNSL (the client).

A significant amount of additional accommodation, provided to generate income, a need arising from the requirement for the stadium to be financially self-sustainable. High design standards required by the lottery funding agreement. Overall capacity of 90,000 seats, which are proportionately more expensive than those in a lower capacity stadium.

(Cyril Sweet, 2002)

Size-related estimating

These techniques are invariably based on gross floor area (GFA) approaches when the total floor area of the required building is calculated and then multiplied by an appropriate unit rate per square metre of floor. In former times volumetric approaches were used, but this technique has largely fallen out of favour as large errors can arise.

More detailed approaches can be applied by the use of differential rates for different areas within the building to give a greater degree of accuracy. A major limitation of these techniques is that they take no account of the geometry of the building.

Elemental cost-analysis estimating

This technique relies on the selection of one or more suitable cost analyses and adjusting them in time, quantity, quality and location in order to provide an estimate of the building. It is a technique which is used as the means to establish the cost plan which should confirm the budget set at the feasibility stage and to establish a suitable cost distribution within the various elements.

Invariably an outline cost plan is first produced using the cost per square metre of each functional element in order to allocate elemental cost limits. When the design has developed further the elemental unit quantities are calculated in order to establish elemental-cost targets for inclusion in the detailed cost plan.

As the design evolves more information becomes available. The element unit rate can be modified as described below. Most elements have different specifications, with varying rates that need to be isolated. For example, a factory unit may have mainly unfurnished warehouse, some offices and a toilet block.

Panel 4.2 The building cost information service of the RICS

The nationally available BCIS contains two types of elemental cost analyses:

1. Concise cost analysis, which gives only a breakdown into broad elements such as substructure, superstructure, services etc.;
2. Detailed cost analysis in a standard format – fully describing each project thus:

    (a) information on the project, including description, site and market conditions, number and prices of tenders, contract period, form of contract;
    (b) elemental costs – showing element total cost, cost per square metre gross floor area, element unit quantity, element unit rate, with two sets of figures for the preliminaries shown separately and apportioned;
    (c) specification and design notes;
    (d) plan and elevation.

The elemental cost data from previous projects is accessible to subscribing members via computer link to the BCIS. Thus even in the absence of designer's drawings, the client's cost adviser is able to create on the computer a pre-contract cost model using the data from several similar previous projects.

Table 4.3 Establishment of elemental cost targets for inclusion in cost plan.

Element 43: floor finishes

Area	Specification	Quantity (m²)	Rate (£)	Cost (£)
Outline cost plan				
All areas	Typical mix	10,000	7.30	73,000
Detailed cost plan				
Warehouse	Floor hardener	9,000	2.00	18,000
Office	Carpet	900	50.00	45,000
Toilets	Ceramic tiles	100	100.00	10,000
		10,000		73,000

The element unit rate calculation arrives at the same cost but assumes an identical mix of specification to arrive at an aggregate rate. It is not easy to fully appreciate an aggregate rate, as it bears no relation to the specification rates. Any change in the ratios of the varying specification could have a significant cost effect. The parts of elements are referred to as components and are added together to create the elemental sum (refer Table 4.3).

Cost checking

In order to confirm the accuracy of the cost plan, which in itself will have confirmed the budget set at the feasibility stage, cost checking is deployed. Cost checking is the execution of the cost-control component in the design process. It ensures that the information as a basis for the tendering can be prepared such that the lowest tender will confidently equate closely with the budget set at the feasibility stage.

Milestone reports

If all the documentation is formatted in the same way it can be compared and reconciled. One way to do this is to use milestone reports. This is a table that summarizes and reconciles between each milestone. A milestone report is normally a report such as

- original budget
- cost plans for example 1 to 10
- pre-tender estimates
- contract sum
- financial statement for example 1 to 50
- final account.

The main-group costs, such as total finishes, are tabled together with the total cost, area, cost per square metre and a comment on any major changes to brief. The costs can be plotted on a graph. After several projects, the client's cost consultant can analyse their performance to see whether they need to adjust their level of optimism or pessimism.

4.5   General comments

The client obviously remembers the first figure reported to them. When the original feasibility study is performed the budget is often fixed; it is therefore essential that all cost reporting reconcile back to the original budget. All estimates should explain to the client and the design team about what is included in the budget. It should be a discussion document for design optimization.

As the design develops it is inevitable that some over-specification in individual elements will occur, sometimes increasing the total cost beyond the total budget. The elemental breakdown can highlight the offending elements by showing an excessive percentage of the total.

Normally the individual percentages of each elemental cost for a particular type of building produce a typical *pattern*. It is important to match the percentages, or pattern, with the norm for that type of building. Mere consideration of the total cost per square metre can be misleading as there can be two high and two low elements, which may cancel each other out and yet require a detailed examination.

4.6   Action after receipt of tenders

In most cases, sound cost planning will produce tenders within budget. If, due to market conditions or late changes in designs and specification, adjustments need to be made to a tender, information on potential savings will need to be identified by the design team. If there are significant changes from the initial tender documents, consideration should be given to the need for seeking revised tenders.

4.7   Conclusion

This chapter has briefly reviewed the key concepts of pre-contract cost management in the industrial engineering, civil engineering and building sectors in order to identify a suitable approach for construction works. Many similarities in the approach and techniques in the three sectors have been identified.

The cost manager's systems should provide clients and project managers with the maximum possible advance warning of likely expenditure so that timely and appropriate actions may be considered.

It is necessary to identify what items are included in the estimate and which are excluded. Forecasts should not be single figures, implying a degree of accuracy that does not exist; they should be a range of figures within stated parameters. Ideally, each estimate should be a logical development of its predecessor reflecting the increased level of detail available.

In the periods between revisions of estimates and cost plans, the development of designs and programmes and the progress of procurement and commitment must be controlled.

A major factor in the management of costs is the identification and management of the risks. Risks are associated with the unknown. Therefore, as a project progresses from inception through design construction and use, the unknown elements should diminish and the risk allowance reduced accordingly.

4.8   Questions

1.   Describe and discuss the range of cost models and show how they are useful at different stages in the design process.
2.   Discuss the steps that can be taken to ensure that cost planning and control keep the final cost of a building project within cost target.
3.   The primary function of producing estimates of the cost of construction works is to be able to advise clients of anticipated development costs. Discuss the various methods of providing such pre-contract estimates in relation to a proposed marina development.

Bibliography

Ashworth, A. (2004) *Cost Studies of Buildings*, 4th edition, Prentice Hall
Bathhurst, P.E. and Butler, D.A. (1980) *Building Cost Control Techniques and Economics*, Heinemann
Ferry, D.J., Brandon, P.S. and Ferry, J.D. (1999) *Cost Planning of Buildings*, 7th edition, Blackwell Science
Flanagan, R. and Tate, B. (1997) *Cost Control in Building Design*, Blackwell Science
Jameson, A. and Nugent, H. (2006) 'Delays ahead as road costs soar', *The Times*, Monday, 23 January 2006, p. 3
Morton, R. and Jaggar, D. (1995) *Design and Economics of Building*, E&FN Spon
Norman, A. (1994) 'Cost Estimating', notes presented at the UMIST Project Management Course, UMIST
Royal Institution of Chartered Surveyors (RICS) (1992) *Cost Management in Engineering Construction Projects: Guidance Notes*, Surveyors Holdings Limited
Royal Institution of Chartered Surveyors (RICS) (1998) *The Surveyors' Construction Handbook*, Surveyors Holdings Limited
Seeley, I.H. (1996) *Building Economics: Appraisal and Control of Building Design, Cost and Efficiency*, Macmillan
Shrimpton, F.S. (1988) 'Cost Management and Reporting in Civil Engineering', RICS/ICE Discussion Meeting, RICS Quantity Surveyors Division
Swinnerton, D. (1995) 'Estimating techniques and their application', *Project*, April, pp. 11–14
The Joint Development Board (1997) *Industrial Engineering Projects: Practice and Procedures for Capital Projects in Engineering, Manufacturing and Process Industries*, E&FN Spon

5 Cost management on PFI projects

5.1 Introduction

Privatized infrastructure projects have been around for at least two hundred years. In the eighteenth century, one of the first concessions to be granted was given to the Perrier brothers to provide drinking water to the city of Paris. The Trans-Siberian Railway and the Suez Canal were thought to have been the first Build-Own-Operate-and-Transfer (BOOT) projects in the modern world (Merna and Njiru, 2002). The great Victorian contractor Weetman Pearson not only built railways, power stations and ports in Britain, Mexico and Chile but also promoted the companies, raised the capital and ran the operations for a number of years.

The idea behind the BOOT philosophy is that there is an increasing worldwide perception that the electorate requires improvements in the quality and availability of its public services, particularly infrastructure projects – but is not prepared to pay extra tax to fund them. A solution is to get the private sector to pay for these facilities (as well as designing, building and operating them) and in return allow the companies involved to take the bulk of the revenue produced.

In recent times the term Build-Operate-Transfer (BOT) was first introduced in the early 1980s by Turkey's then Prime Minister, Turgat Ozal. Under this arrangement the private organizations undertake to build and operate a project normally undertaken by government. Ownership normally reverts to government after a fixed concession period, normally between 10 and 50 years. The revenues generated by the project are the main source of repaying the debt. The projects are normally structured to have limited or no recourse to the project sponsors, contractors or to the government – projects being undertaken by a self-contained concession company or special-purpose vehicle.

This approach is particularly attractive for governments in the rapidly developing countries in the Pacific Rim, for example, in Thailand and Malaysia, which see the BOT approach as a means of reducing public sector borrowing, and at the same time promoting direct foreign investment in their country's infrastructure or industrial projects. Examples of such projects include power stations, toll roads, toll bridges and even pipeline systems for oil and gas.

The proactive involvement of the home government is usually critical to the success of the project. Robert Tiong's important research in the early 1990s identified that sponsoring governments adopted a range of strategies to support these major infrastructure projects. These included giving concession periods up to 55 years, offering support loans, giving concessions to operate existing facilities, facilitating foreign exchange guarantees and interest rate guarantees etc. (Tiong, 1990).

The first BOT project in the UK was the Channel Tunnel linking UK and France, constructed by a consortia of five British contractors, five French contractors and five banks. This was followed by other infrastructure projects including the Skye Bridge, the Second Severn Crossing, the Dartford Bridge, the London City Airport and the Manchester Metrolink.

5.2   Structure of BOT projects

The main parties involved in a BOT project are as follows:

- *The host government*: often the host government provides critical financial support without which the project would not become a reality;
- *The project sponsors*: normally a joint venture comprising contractors/ banks/entrepreneurs;
- *The banks*: may include major world banks, for example Asian Development Bank, European Investment Bank as well as major national banks;
- *The shareholders*: include pension-fund holders and major investors;
- *The contractors*: often multinational joint ventures.

The typical five phases of a BOT project, with the roles and responsibilities of the project sponsors identified, are as follows:

1. *Pre-investment*: acting as consultants the project sponsors carry out the feasibility study;
2. *Implementation*: acting as consultants the project sponsors carry out the engineering/ building design, as project sponsors they negotiate favourable concession agreements from government, and as project promoters they raise equity and borrow finance during the implementation phase;
3. *Construction*: the project sponsors act as the contractor to build the facility, usually on a fixed-price turnkey basis, during the construction phase;
4. *Operation*: the project sponsors act as the operator and owner of the facility, using the project revenues to repay the loans during the operation phase;
5. *Transfer*: transfer of ownership to the government.

5.3   Case study: Nottingham Express Transit (NET) Light Rail

In 1990 the Nottingham City Council and Nottinghamshire County Council formed a private company limited by shares under the name of Greater Nottingham Rapid Transit (GNRT) to undertake the construction and operation of the LRT system.

In 1994 the Act of Parliament (GNLRT Act) was passed granting the City and County Councils the powers to develop and operate an LRT system, to authorize the construction of the works and acquisition of land required and to transfer the undertaking to GNRT Limited or any other organization.

In 1997 Arrow Light Rail was selected as the preferred bidder. Arrow Light Rail is an SPV company formed to design, build, fund, operate and maintain Line One of the Nottingham Express Transit.

The promoter is Nottingham Express Transit (NET), the private company formed by Nottingham City Council and Nottingham County Council (GNRT under the Parliamentary Act).

The concessionaire is Arrow Light Rail comprising six partners each bringing their own particular skills and expertise to the organization.

1. *Bombardier (formerly Adtranz)*: one of the leading providers of total rail systems and tailor-made solutions for rail-transport services worldwide;

2. *Carillion Private Finance (CPF)*: part of Carillion plc, the leading services to construction company. CPF is a UK leader in private finance, primarily in health, prisons and transport sectors;
3. *Transdev*: the leading operator of integrated urban transport systems in Europe and operating over 70 public transport networks in France including modern tramways;
4. *Nottingham City Transport*: the leading bus-operating company in the Nottingham area.
5. *Innisfree*: the leading private equity investor in the UK Private Finance Initiative (PFI) and Public–Private Partnership (PPP) infrastructure projects and a 30% shareholder of Arrow;
6. *CDC Projects*: a 20% shareholder of Arrow, the major French public sector financial institution.

Figure 5.1 shows the PFI organizational structure on stage 1 of the Nottingham Tramway. The contractor is Bombadier Carillion Consortium (BCC), a segregated consortium in which the two members take responsibility for delivery of their respective scopes of work, and are jointly and severally bound to supply the entire scope of works to the client – Arrow Light Rail.

The operator is the Nottingham Tram Company, a special-purpose company owned by Transdev and NCT, which took over the project on completion.

Arrow's £220 million funding of the project is met by bank loans, sponsor equity and grants. The loans will be paid back during the operating period from performance-related payments from the promoters and revenue from fares.

The promoter awards a concession to the concessionaire who in turn procures design and construction from the contractor, and operations from the operator. The principal documents are as follows:

- *Concession agreement and schedules*: specifies the rights and obligations of the promoter and the concessionaire;
- *Turnkey contract and schedules*: specifies the rights and obligations of the concessionaire and the contractor;
- *Railtrack agreements*: specifies the rights and obligations of the promoter, the concessionaire, the contractor and railtrack for the railtrack-enabling works.

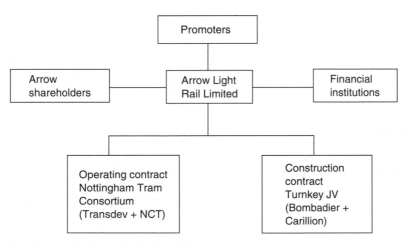

5.1 PFI organization on Nottingham Tramway.

Both members of the Bombadier Carillion Consortium are bound by the Consortium Agreement, which defines their respective scopes of work and responsibilities within the Consortium.

Some ten years in the planning, NET is a key part of Greater Nottingham's integrated transport strategy, to improve access to the city centre and reduce traffic congestion and pollution. Line 1 was anticipated to take 2 million car journeys per year off the roads and was also expected to bring economic regeneration and environmental benefits to the area. The project finished six months late, the public sector was protected by a robust PFI structure; however, the turnkey contractors, mainly civil engineering lost money on the project.

NET Line One was opened by Transport Minister Alistair Darling in March 2004 and has shown strong patronage with 8.4 and 9.7 million passengers respectively in its first two years of operation. A recent passenger survey revealed 98% satisfaction with the service and 80% support for the planned new lines south and west of the city.

5.4   Factors leading to success on BOT projects

The following factors have been identified by the United Nations Development Unit as important for the success of BOT projects (United Nations, 1996):

1. The project must be financially sound, feasible and affordable.
2. The country risks must be manageable.
3. There must be strong government support.
4. The project must rank high on the host government's list of infrastructure projects.
5. The legal framework must be stable.
6. The administrative framework must be efficient.
7. The bidding procedures must be fair and transparent.
8. BOT transactions should be structured so as to be concludable within a reasonable time and at a reasonable cost.
9. The sponsors must be experienced and reliable.
10. The sponsors must have sufficient financial strength.
11. The construction contractor must have sufficient experience and resources.
12. The project risks must be allocated rationally among the parties.
13. The financial structure must provide the lenders adequate security.
14. The foreign exchange and inflation issues must be solved.
15. The BOT contractual framework must be coordinated and must reflect the basic economics of the project.
16. The public and the private sectors need to cooperate on a win–win basis.

Building a BOT toll road in a developed country might be an attractive proposition to a potential sponsor. However, in contrast, building such a road in a developing country might not be so attractive due to the economic and political situation and potential risks involved.

5.5   Risks and securities

The United Nations Development Unit (United Nations, 1996) has identified the range of potential risks on major international BOT projects which could include the following:

- Political risks including changes in taxation, import restrictions or even cancellation of the concession;
- Country's commercial risks including foreign exchange, inflation and interest rates;
- Country's legal risks including changes in law and regulations;

- Development risks including bidding, planning and approval;
- Construction risks including delay, cost overrun, force majeure, loss or damage to work and liability;
- Operating risks including technical, demand, supply, cost escalation and management risks.

The Channel Tunnel project highlighted the risks to investors involved in BOT projects. The project opened more than one year late and cost at least three times the initial budget of £3.5 billion. In contrast, the £80 million QE2 bridge on the Thames crossing at Dartford was completed on time and within budget and will show a generous return for the investors; it is anticipated that the toll revenue in the last year before handover to the government will equate to the total initial cost.

Given the complexity of BOT schemes and the magnitude of the funds required, it is important for project sponsors to identify, evaluate and allocate the main risks in a BOT project. Indeed, the analysis and allocation of risk is central to the financial structuring of a BOT project. In the first instance, the objective must be to minimize the risks associated with the project, for example, by adequate geological, technical and market studies. Thereafter, the process is one of insuring, controlling and apportioning risks according to the parties' willingness to bear them.

The key to successful BOT project financing is structuring the project finance with as little recourse as possible to the sponsors or government, while providing sufficient guarantees and undertakings so that lenders will be satisfied with the credit risk.

PFI projects have significant benefits for clients, not only is the project considered off the balance sheet but all the risks should be transferred to others. However, the termination of the PFI contract for the National Physical Laboratory (NPL) demonstrated that sometimes the risks may be too great for the private sector to carry.

In 1998 Laser, an SPV jointly owned by Serco Group plc and John Laing plc, signed a 25-year PFI contract under which Laser would build and manage new facilities for NPL, comprising over 400 laboratories. Laser awarded John Laing a fixed-price contract to design and build the new facilities. However, intractable problems occurred in designing 30 specialist laboratories which required stringent temperature and/or the sub-audible noise controls facilities which considerably delayed the project.

Ultimately John Laing plc was unable to carry the massive financial burden, with losses estimated at £67 million; John Laing Construction was sold and the remaining company restructured. In turn Serco recognized that it could not complete the project and the PFI agreement was terminated in 2004 (www.nao.org.uk/publications [accessed 20 May 2007]).

Postscript: The project was completed by the client, the Department of Trade and Industry, in 2007.

5.6 Case study: Sydney SuperDome, Australia

In the last two decades the BOOT strategy has become popular in Australia with examples such as the Sydney Harbour Tunnel, the Stadium Australia in Sydney, the M2, M4 and M5 toll ways in New South Wales and the Ord River Hydro-Electric Scheme in Western Australia.

The Sydney SuperDome is a 70,000 m^2 multi-use indoor arena with seating capacity for 20,000 spectators. It was built at a cost of A$280 million (£115 million) over 25 months as part of the 2000 Olympic Games infrastructure programme with a 30-year operation concession period. The SuperDome, which is the largest indoor venue in Australia, can be used for gymnastics, tennis, basketball and ice hockey and also for concerts and exhibitions. The SuperDome was designed and constructed by Abigroup together with the Japanese big-five contractor Obayashi. Figure 5.2 shows diagrammatically the simplified corporate structure clearly identifying the main parties in the project.

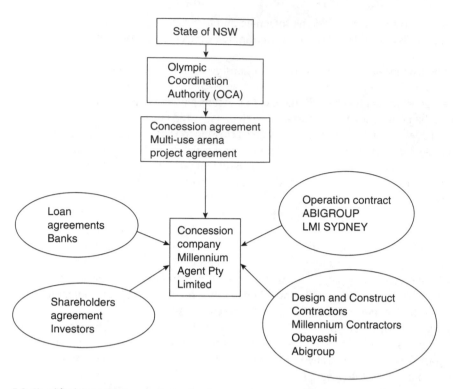

5.2 Simplified SuperDome corporate structure.

The Australian government contributed 72% towards the cost of the arena alone with the Millennium Consortium contributing the balance. The Australian government also paid the Consortium for the construction of an adjacent car park and landscaping.

The Multi-use Arena Project Agreement obliged Millennium Agent to

- finance, plan, design, construct and commission the arena;
- plan, design, construct and commission the adjacent car park and public domain areas;
- procure the operation, maintenance and repair of the arena during the lease until 2031;
- make the arena available for the Olympic Games;
- yield up possession of the arena to the OCA in 2031 (Jefferies, 2006).

The Sydney SuperDome is a success story and is setting the benchmark for future PPPs to be initiated by the Australian New South Wales Government. In 2006 it was renamed the Acer Arena (www.acerarena.com.au [accessed 24 March 2007]).

5.7   The Private Finance Initiative (PFI)

The PFI was launched in November 1992 by the then chancellor of the exchequer, Mr Norman Lamont. It came against the backdrop of a major programme of privatization during the 1980s under the leadership of the then prime minister, Margaret Thatcher.

The PFI is a form of PPP that marries a public procurement programme, where the public sector purchases capital items from the private sector, to an extension of contracting-out, where public services are contracted from the private sector. Under the PFI structure revenue is received from the government. In contrast, PPP projects are stand-alone and receive no government support.

In 1994, the then Conservative government introduced the requirement for all central ministry departments to check whether every project they planned was capable of being procured under PFI. With the change of government in May 1997 the Labour government conducted a thorough review of the experience of PFI in the previous five years. The results of this review – the Bates Review – endorsed the use of PFI and led to a substantially improved process and simplified market (CIC, 1998).

Since 1996, PFI has accounted for around 10–15% of all the UK's gross public investment with the health and education sectors contributing the greatest number of projects. However, the largest capital-value projects are in the transport and defence sectors with the £10 billion Channel Tunnel completed in 1994 and the £17 billion maintenance and upgrading of two-thirds of the London Underground by Metronet.

In July 2007 Metronet went into administration claiming that the 30-year project had already overspent by £1.2 billion in the first seven-and-a-half years due to London Underground repeatedly asking for work outside the original scope. Metronet's five shareholders WS Atkins, Balfour Beatty, EDF Energy, Bombardier and Thames Water had already written off considerable sums from their company balance sheets. This case study clearly identifies that major PPP/PFI projects involve potential massive risks for private companies and their shareholders and may not be appropriate for all public projects.

As of December 2006, 794 PFI deals had been awarded by the UK Government including expenditure on schools replacement and refurbishment, prisons, hospitals, new or refurbished health facilities, transport projects including roads bridges, waste and water projects with over 150 local authorities involved in PFI projects (CBI, 2007).

The UK Government is committed to using PFI as a procurement option wherever it is value for money to do so. The HM Treasury paper *PFI: Strengthening Long-term Partnerships* (HM Treasury, 2006) identified that PFI has a strong record on delivering projects on time (76% compared to 30% on non-PFI projects) and on budget (79% compared to 27%).

Five key lessons concerning PFI projects have been identified in the relentless search for improvement in public services (CBI, 2007):

1. Integrating service and design leads to a whole-life approach;
2. Allocating risks creates incentives for better delivery;
3. Increasing transparency and accountability has wider benefits for government financing;
4. Improving customer and staff satisfaction is key;
5. Opportunities are created to prioritize environmental sustainability.

However, PFI projects are not without controversy with two principal complaints. First, the high cost of bidding (estimated at £11.5 million for hospitals and £2.4 million on schools projects), and second the diminishing number of firms who are prepared to do so.

Structure of PFI projects

The aim of the PFI initiative is to bring private sector skills into projects that would have previously been wholly or mainly provided by the public sector. The underlying principle behind PFI has many dimensions. The obvious one is the pure PFI case where a facility and service is provided at minimal cost to the public sector. Second, it is the public sector's exploitation of the private sector's ability to design and manage more efficiently. The public sector is characterized by substantial cost overruns and poor management skills; utilizing PFI and passing over control may eliminate some of these inefficiencies.

The choice of PFI as the most appropriate route for procurement is governed by Treasury rules for developing an outline business case. The government department has to support an application for funding with appropriate feasibility studies. In addition, the department has to demonstrate that significant benefit would be derived from this type of contract over any alternatives. This includes cost savings, but is principally governed by the benefits of transferring inherent risks to the private sector.

In PFI accommodation projects, such as hospitals or prisons, the construction element typically represents around 25–30% of the total value of the contract. But other project costs, such as maintenance, will be influenced by the quality of the construction work.

Figure 5.3 shows the commercial structure of a PFI project with the three main parties: the government customer, the operation carried out using a special-purpose vehicle (SPV) and the provider of finance.

The key role in any PFI project is an SPV which is the legal entity; essentially a shell organization which is created by shareholders in a bidding company or consortium. The SPV is designed to fund the project and contract with the public sector client for the purpose of delivering the PFI service. Depending on what is included in the contract, the SPV is likely to include the equity provider, the contractor and a facilities management provider. During the selection process the SPV will need to provide detailed financial models showing how, amongst other things, they intend to fund a project over the lifetime of the project.

The concessionaire is the company, which has been awarded the contract by the public sector client to provide the PFI service. The concessionaire company will either be an SPV established by a bidding firm for the sole purpose of the specific project, or an existing company, which will be taking the liabilities of the PFI, project 'on balance sheet'. The concessionaire is responsible for the design, construction, financing and operation of the built facility required to provide the PFI service.

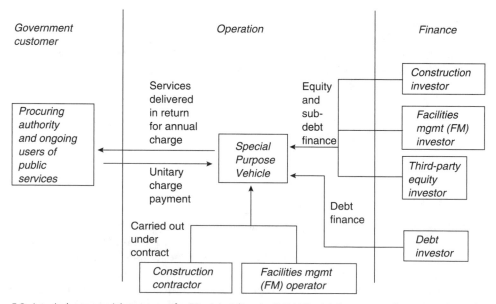

5.3 A typical commercial structure of a PFI project (source: World Bank Infrastructure Governance Roundtable United Kingdom PPP Forum (www.worldbank.org/transport/ learning [accessed 23 March 2007])).

Types of PFI strategies

The Treasury has described three different types of PFI strategies:

Financially freestanding projects

1. In these projects the private sector undertakes the design, building, finance, operation and maintenance of the completed asset. The project may involve building a new asset or taking over the operation and maintenance of an asset.
2. Examples include the Skye Bridge, the A50 Stoke-Derby link road and the M40 widening. The concessionaire's revenue is generated from the collection of tolls (real or shadow – based on monitored usage).

Services sold to the public sector

In this type of PFI project, the cost of the project is met wholly or mainly by charges from the private sector service provider to the public sector body, which let the contract. Examples include privately financed prisons and hospitals. In the case of prisons, the Prison Service does not lease the prison accommodation; rather it pays for a complete service provided to the inmates.

Joint ventures

The cost of joint ventures is met partly from public funds and partly from private funds, with overall control of the project resting with the private sector. Examples of such projects include the Channel Tunnel Rail Link and the Croydon Tramlink.

This type of project requires a value-for-money test conforming to the following criteria:

*   Private sector partner chosen in competition;
*   Joint-venture control held by private sector;
*   A clear definition of government contribution and its limitations;
*   Clear agreement on risk and reward allocation, defined and agreed in advance.

Government contributions can take various forms and include equity, concession loans and asset transfer. Typically, government contributes through aiding initial planning or granting subsidies.

The PFI Process

The four 'p's (4ps) (Public–Private Partnership Programme) is a local government central body set up to assist local authorities to develop and procure projects through Public–Private Partnerships and the PFI. The 4ps (pronounced four-peas) offers comprehensive procurement support to local authorities including hands-on project support, gateway reviews and skills development and know-how procurement guidance in the form of procurement packs, case studies and extranets.

The 4ps publication *A Map of the PFI Process: Using the Competitive Dialogue* provides an overview of the process as a whole and a short introduction to each of the identified stages (www. 4ps.gov.uk [accessed 23 March 2007]). The guide states that the PFI transforms local authorities from being the owners and operators of assets, to the purchasers of services. In a PFI transaction, a private sector service provider is given responsibility for designing, building, financing and operating assets, from which a public service is delivered.

Once the need for the service/facility is identified, procurement options are considered and investment appraisal exercises undertaken. The conclusions of the detailed options appraisal exercises should be documented in an Outline Business Case (OBC).

From the appraisal of a number of possible project and procurement options, a preferred service delivery option the *Reference Project* will be identified; this is the benchmark solution against which bids are subsequently identified. The same project option procured through a traditional route is known as the Public Sector Comparator (PSC) and provides a benchmark against which the value for money (VFM) of the PFI can be assessed at the OBC stage.

The output specification, developed as part of the suite of bid documents provided to bidders, is arguably the most important document in the procurement of a PFI project. It sets out what is expected of the SPV, written in terms of outputs or outcomes. This means specifying what the authority expects to see as a result of the project rather than how it expects that result to be achieved thus encouraging innovation.

The payment mechanism has two main elements: availability standards and performance standards. The availability standards define when the asset is considered *available for use*. The performance standards cover any standards not covered by the availability standards, in general the service and maintenance standards.

The authority pays a monthly or yearly *unitary charge* to the SPV. The charge is calculated based on the availability and performance of the facilities and associated services and deductions are made for poor performance. In order to determine the unitary charge prior to submitting their tender for the project, the sponsors would have to estimate the following:

- The capital costs of the project;
- The likely financing costs, and hence, debt-service responsibilities of the contractor;
- The operating costs of the project for 25–30 years (including subcontractor costs, administrative costs, employment costs, insurance costs, tax liabilities and other costs, expenses and fees;
- Any other costs, expenses and risks inherent in the project;
- Surplus by way of dividends or interest on subordinated loans.

At the end of the contract term the Authority might have the right to acquire the building at a specified value or to walk away if it has no further need for the asset. Alternatively, there might be provision to re-tender the service with the building being made available to the successful bidder. Whether the SPV get paid for the asset will depend on the approach to residual value and whether this is reflected in the charges.

Contract management under PFI

The invitation to tender should include a draft of the PFI contract – called the model contract. This sets out the terms on which the local authority expects bidders to submit a standard bid. The HM Treasury *Standardisation of PFI Contracts* and the service-specific model contracts included in 4ps *Procurement Packs* (where relevant) should be used to develop the model contract.

Contract management is the process of managing and administrating the PFI/PPP contract from the time it has been agreed at contract award, through to the end of the concession period. The 4ps publication *A Guide to Contract Management for PFI and PPP projects* (www.4ps.gov.uk [accessed 23 March 2007]) identifies that there are four components of contract management:

1. *Setting up the contract management team*: determines when the contract management should be set up, the structure, the attributes and training needs.
2. *Managing relationships*: establishes relationships, communication routes and systems, and the active support and enhancement of them throughout the life of the project.

3. *Managing service performance*: assesses whether the services being delivered by the service provider meet the required standards and whether remedial measures are effective.
4. *Contract administration*: ensures obligations and responsibilities defined under the contract are met, ensuring underperformance, risks, payment of the unitary charge, reporting and change are all managed effectively so that VFM and continuous improvement are achieved.

The 4ps guide also includes a map of contract management involvement throughout the project development and procurement stages.

5.8   The role of the cost consultant in PFI/PPP projects

The cost consultant has a significant role to play under the three procurement systems: traditional, design and build, and the management approach. Their expertise has developed based on a sound appreciation of construction technology, measurement and estimating, legal and contractual, budgeting, cost-planning and cost-control systems, change and claims management. A challenge for the profession is to provide services, which add value under the PFI procurement route.

During the early stages of the project the client may rely on external advisors to supplement the skills and knowledge bases of the in-house personnel. This is particularly relevant if the client is new to PFI or if in-house resources are being utilized on other projects.

There are three main categories of professionals involved in PFI/PPP projects: financial, legal and technical. The financial advice is usually undertaken by personnel from the major banks and specialist PFI advisors. Likewise, legal advice is usually undertaken by leading lawyers. The technical team includes traditional construction-related specialisms and facilities management disciplines.

The process of appointing external consultants will be dependent upon public sector procurement procedures appropriate to the market sector. Clients traditionally appoint advisers independently, but a turnkey appointment may also be used, as consultants gain experience of working effectively with each other.

The structure of a PFI project is complex and involves many different parties. Eaton and Akbiyikli (2004) identify that this should create the following advantages for quantity surveyors (QS): QS services are required by more parties; added value services are now required and QS services are needed for longer duration. These QS services are required by a potentially wider range of clients including the public sponsor, the SPV, the design and construction companies, the operation and maintenance companies, and by the various financial institutions providing equity, commercial loans and debt finance.

The structure for the construction phase of a contract is typically representative of traditional procurement – usually based on the design and build approach. The operational phase will reflect typical facilities management contract structures.

In addition, cost managers have established themselves as independent certifiers – crucially being involved at the sign-off of the project for occupation or use by the client body. Table 5.1 indicates the range of roles undertaken by construction cost and project managers on PFI/PPP projects.

Under the PFI structure the QS needs to develop expertise in relatively new areas of knowledge including checking the sustainability of the design, validating the acceptability of innovative solutions, developing more expertise in risk management, developing payment systems linked to performance measurement, executing audit procedures and undertaking life cycle and operational cost management.

Table 5.1 Roles of cost manager and project manager on BOT/PFI projects.

Client		Cost management	Project management
Public sector	Project sponsor	Construction cost advice	Bid process management
		Public sector rules/OGC gateway advisory and review	OJEC procedures
		Business case, public sector comparator advice	Advisor team management
		FM cost advice	Construction advice
		Procurement strategy	Long-term contract management
		Commercial negotiations	
		Tender analysis	
		Risk advice	
		Capital tax advice/VAT	
		Life cycle cost advice	
		Output specification development	
		Technical evaluation	
Independent	Sponsor and bidder	Independent certifier	Independent certifier
Private sector	Bidder	Construction cost manager	PM for SPV
		FM cost manager	SPV bid manager
		Risk manager	Construction cost manager
		Capital tax advisor	
		FM consultancy	
		Life cycle cost manager	
	Financier	Technical advisor	Technical advisor

Source: Adapted from Hedgecox, D. *Costing out PFI/PPP* (www.publicservice.co.uk/pdf/pfi/spring2001/p168.pdf [accessed 27 March 2007]).

5.9   Case study: Stoke-on-Trent Schools, UK

In 1997 many of the schools in Stoke-on-Trent were in a dilapidated state and not fit for modern teaching and learning practice. The schools included buildings dating back to the nineteenth century, some of which had not been upgraded or refurbished for 50 years. Furthermore, the annual budget for maintenance of all the schools was £120,000; totally inadequate when one large replacement boiler would cost £80,000. The City Council's annual expenditure was already stretched to its absolute limit so a radical brave new way of thinking was required.

In November 2000, after three years of intense planning and negotiation, one of the first PFI partnership charters in the UK was signed to cover the refurbishment and maintenance for 25 years of all Stoke-on-Trent's 122 schools. This five-year capital expenditure scheme was very much a pioneer in PFI school projects. There was no precedence to follow and no standard contracts were available at the time. It is noted that the Treasury Task Force later issued guidance for standardizing the terms of PFI contracts.

The project cycle followed the 4ps process with independent *Gateway Reviews* at key points, all as the OGC *Gateway Process* model (www.ogc.org.uk). The different procurement options were considered before selecting the PFI approach. Feasibility studies were undertaken using the public sector comparators as the benchmark. Output specifications were developed embracing such issues as minimum temperature in classrooms. Sophisticated financial models, which included sinking funds and risk allowances, were developed and rigorously tested.

The winning bid was received from a special-purpose company called Transform Schools (Stoke) comprising shareholders Balfour Beatty Capital Projects (50%) and Innisfree (50%). The project is unique in being the largest bundled refurbishment scheme ever attempted in England and is valued at £153 million of which £80 million is for building nine new schools and refurbishing the rest of the portfolio. One of the principal reasons that Balfour Beatty succeeded in securing the contract was its innovative proposal to replace nine schools rather than refurbish them.

Once the SPV was chosen there was a 12-month intense period of activity in which the architects Aedas worked with the school governors to develop acceptable designs. The construction contracts were let on a design and build basis.

The PFI Board, comprising City Counsellors, representatives from the Department for Education and Skills (DfES) and 4ps together with the authority's project director, met on a monthly basis. The PFI team comprising the authority's project director, lawyers, financiers and the technical team met on a two-weekly basis. Some meetings comprised over 40 participants so one of the biggest challenges facing the project team was capturing the knowledge and expertise and incorporating the feedback into the project.

The key lessons learned from this project include the following:

- A real belief in the partnering ethos by all the parties. There were difficult problems to resolve throughout the negotiation period but the parties kept talking until these were finally resolved. The original contract, with 9 new schools, was extended to a total of 17 new schools; 15 were built by the SPV contractor and 2 by other contractors.
- Detailed identification and evaluation of the main risks throughout the 25-year period to be passed to the SPV. In this contract, the additional risks were estimated at 17%. Some risks were considered unreasonable for the SPV to carry and were retained by the authority, for example vandalism in school time.
- The importance of teamwork with the complete integration of the key stakeholders in an open forum.
- Attention to detail in the innovative contract which included

  a. a clause requiring the contractor to demonstrate a 20% saving in energy consumption in each school in the first five years by 2006 and a further 5% saving by 2010;
  b. an agreement on the refinancing provision with the authority retaining 25% of any profits; this was a particularly difficult point for the negotiation team with the SPV wanting to retain the whole benefit, while the authority wished to take a 50:50 split;
  c. the client's involvement in securing a quality design, for example, they could comment at the point of handover and the contractor might be required to make changes at their own expense if not acceptable;
  d. a stipulation that at the end of the 25-year period the estate should be in a position where there would be no major repair necessary for the next five years;
  e. change or variation clauses allowing the authority to bring in other contractors to do the work if the SPV contractor's price for the variation was considered too high.

This pioneering PFI project not only provided one of the best portfolios of school buildings in England but also resulted in other positive features including employment of 500 local labour during construction; apprentices taken on by the SPV contractors and sponsorship of a local community football team. Most significantly, there has been a dramatic reduction in school vandalism and a raising of student and teaching staff morale. It is anticipated that the improved buildings will also result in improved student performance in the years to come.

The main parties involved in this pioneering project were as follows:

Client: Financial Advisor: ABROS; Legal Advisor: Eversheds; Technical Advisors: Gleeds and the Building Research Establishment;
SPV: Financial Advisor: HSBC Plc; Legal Advisor: Clifford Chance; Legal Advisors: Clifford Chance, Tods Murray; Technical advisors: Summers Inman (life cycle assessments, technical

assessments), Hurst Setter (health and safety), Capita (M&E and structural), Walker Cotter (planning supervisor) and Atkins Faithfull & Gould (monitoring engineer/technical advisor to the lending banks);

Service Provider: Transform Schools (Stoke) Limited; Shareholders (providing the equity): Balfour Beatty Capital Projects 50% Innisfree 50%; Funders (providing the senior debt funding): Lloyds TSB and Dexia Public Finance Bank;

Subcontractor 1: Stoke Schools JV comprising Balfour Kilpatrick Limited (Design and Build) and Balfour Beatty (Design and Build); Subcontractor 2: Haden Building Management Limited (Hard FM).

Sources: www.transformschools.co.uk (accessed 13 July 2007); www.balfourbeatty.com (accessed 23 March 2007); interview with the Authority's Project Director Mike Inman, 12 July 2007.

5.10   Conclusion

There is strong evidence that PFI projects are far better at keeping to time and budget than other forms of procurement. With support from both the Labour and Conservative parties, it is clear that PFI is here to stay in the UK. The PFI arrangement is attractive to governments worldwide as it allows them to keep public borrowing down as the projects are funded throughout their usable life cycle. However, the evidence on quality of service and to some extent design is mixed.

To date, it would appear that the most successful projects have been of the financially freestanding variety, for example the Dartford crossing. However, there was widespread frustration at the lack of projects financially closed, particularly in the health sector. Likewise, in the highway sector there has been frustration at the time and cost involved in the Design-Build-Finance-and-Operate (DBFO) market. It is not uncommon for DBFOs to take nearly two years from receipt of tender documents to financial close and for the bidding process to cost more than £3 million.

Initially, local authorities had an ambivalent attitude towards the initiative but are now showing interest particularly for the provision of schools, which should create opportunities for smaller contractors executing the building work as subcontractors.

In practice, construction cost managers and project managers have established themselves as key participants within PFI/PPP projects in a number of areas. As well as performing advisory roles for public sector clients, cost and project managers are called upon to assist with the other key parties to a PFI/PPP transaction, namely the bidders and the financiers. Construction cost managers and project managers have a significant role to play in this challenging environment not only in the UK but also internationally in countries such as South Africa and Australia.

5.11   Questions

1.   The proposal is to widen the M6 motorway between Birmingham and Manchester – a distance of 80 miles – from 3 lanes each way to 5 lanes each way.
        Bids are invited from international consortia to execute the project under a PFI arrangement.
        Identify the main risks for the public sponsors, the SPV, the lenders and the contractors under the following headings:

   a.   Social
   b.   Legal
   c.   Economic
   d.   Environmental

e.   Political
f.   Technological.

For solution see Fig. 6.2 p. 68 in David Eaton's *A Report on PFI and the Delivery of Public Services Quantifying Quality*, RICS (2005) www.rics.org/NR/rdonlyres/A37E17C2-C4FC-4F50-AE40– 14F62E949BD3/0/PFI_Report.pdf (accessed 23 March 2007)

2.   Compare and contrast the key differences between traditional public sector procurement and PFI (for a solution see CIC, 1998, p. 10);
3.   Identify the key issues and the lessons to be learned on the PFI case study, *The Redevelopment of the Cruciform Building*, University College London www.hefce.ac.uk/Pubs/hefce/1999/99_43.htm (accessed 23 March 2007)
4.   The NAO has produced 58 reports on PFI/PPP projects. Make a 10-minute presentation to your peers on the key findings and recommendations of one report. http://nao.gov.uk/guidance/pfi_ppp_reports_by_sector.htm (accessed 15 February 2007)
5.   David Eaton's research at the University of Salford, 'A report on PFI and the Delivery of Public Services Quantifying Quality' contains eight case studies. Make a 10-minute presentation to your peers on the key findings and recommendations of one case study. (www.rics.org/NR/rdonlyres/A37E17C2-C4FC-4F50-AE40-14F62E949BD3/0/PFI_Report.pdf [accessed 23 March 2007])
6.   The Commission for Architecture and the Built Environment (CABE), *Client Guide Achieving Well-designed Schools through PFI*, gives a good overview of the design process under a PFI project. Make a 10-minute presentation to your peers on the recommendations from one chapter (chapters 2 to 7) www.cabe.org.uk/AssetLibrary/1919.pdf (accessed 23 March 2007)

Bibliography

Akintoye, A., Beck, M. and Hardcastle, C. (eds) (2003) *Public-private Partnerships: Managing Risks and Opportunities*, Blackwell Publishing

Boussabaine, A. (2006) *Cost Planning of PFI and PPP Building Projects*, Taylor & Francis

Construction Industry Council (1998) *Constructors' Key Guide to PFI*, Thomas Telford

Dinesen, B. and Thompson, J. (2003) 'PFI/PPP Projects – Are They Working?', An overview of the Major Projects Association's (MPA) 21st Annual Conference

Eaton, D. and Akbiyikli, R. (2005) *A Report on PFI and the delivery of Public Services: Quantifying quality*, RICS

Grimsey, D. and Graham, R. (1997) 'PFI in the NHS', *Engineering, Construction and Architectural Management*, vol. 4, no. 3, pp. 215–231

Hawash, K. and Barnes, M. (1997) 'The potential for adopting the NEC's family of contracts on PFI projects', *Engineering, Construction and Architectural Management*, vol. 4, no. 3, pp. 203–214

Hickman, D. (2000) *PFI and Construction Contracts*, Chandos Publishing

HM Treasury (2006) *PFI: Strengthening Long-term Partnerships*, HMSO, available online at: www.hm-treasury.gov.uk/media/1E1/33/bud06_pfi_618.pdf (accessed 6 December 2006)

Jefferies, M. (2006) 'Critical success factors of public private sector partnerships: A case study of the Sydney SuperDome', *Engineering, Construction and Architectural Management*, vol. 13, no. 5, pp. 451–462

Jefferies, M., Gameson, R. and Rowlinson, S. (2002) 'Critical success factors of the BOOT procurement system: Reflections from the Stadium Australia case study', *Engineering, Construction and Architectural Management*, vol. 9, no. 4, pp. 352–361

Kumaraswamy, M.M. and Zhang, X.Q. (2001) 'Governmental role in BOT-led infrastructure development', *International Journal of Project Management*, pp. 195–205

Merna, T. and Njiru, C. (eds) (2002) *Financing Infrastructure Projects*, Thomas Telford

National Audit Office (2003) 'PFI: Construction Performance', Report by the Comptroller and Auditor General HC371 Session 2002–2003, 5 February

Owen, G. and Merna, A. (1997) 'The Private Finance Initiative', *Engineering, Construction and Architectural Management*, vol. 4, no. 3, pp. 161–162

Payne, H. (1997) 'Key legal issues in projects procured under the Private Finance Initiative', *Engineering, Construction and Architectural Management*, vol. 4, no. 3, pp. 195–202

Royal Institution of Chartered Surveyors RICS (2003) *Project Management and the Private Finance Initiative*, RICS, available online at: www.observatory.gr/files/meletes/Project%20management%20and%20the%20PFI.pdf (accessed 23 March 2007)

Tiong, R.L.K. (1990) 'Comparative study of BOT projects', American Society of Civil Engineers (ASCE), *Journal of Management and Engineering*, vol. 6, no. 1, January, pp. 107–122

United Nations Industrial Development Organisation (UNIDO) (1996) *Guidelines for Infrastructure Development through Build-Operate-Transfer (BOT) Projects*, UNIDO, Vienna

Walker, C. and Smith, A. (1995) *Privatized Infrastructure: The Build, Operate and Transfer Approach*, Thomas Telford

Useful websites

www.4ps.gov.uk
http://en.wikipedia.org/wiki/Private_Finance_Initiative

6 Contractor's estimating and tendering

6.1 Introduction

In the *Code of Estimating Practice* the Chartered Institute of Building defines estimating as 'the technical process of predicting costs of construction' and tendering as 'a separate and subsequent commercial function based on the estimate' (CIOB, 1997). At first sight the production of an estimate might appear to be a precise technical and analytical process. However, in reality it is often a subjective process based on knowledge and experience of the key participants.

The estimating process is very important, as it enables construction companies to determine their direct costs, and provides a *bottom line* cost below which it would not be economical for them to carry out the work (Smith, 1999). Overestimated costs result in a higher tender price and rejection by the client. Likewise, an underestimated cost could lead to a situation where a contractor incurs losses. If the contractor is selected then the estimate should also provide the basis for project budgeting and control.

The submission of successful tenders is obviously crucial to the very existence of contractors. Yet, a fundamental truth of competitive tendering, particularly on major works, is that the lowest tenderer is often the one who has most seriously underestimated the risks which obviously could have drastic consequences particularly in times of recession when margins are slim to say the least.

A study by Al-Harbi *et al.* (1994) identified that the main problems facing estimators in Saudi Arabia while compiling tenders for building works included tough competition, short contract period, incomplete drawings and specification, incomplete project scope definition, unforeseeable changes in material prices, changes in owner's requirements, current workload, errors in judgement, inadequate production time data, lack of historical data for similar jobs and lack of experience of similar projects. These items indicate the challenges faced by estimators no matter where the work is carried out.

Further research on the factors influencing project-cost-estimating practice in the UK (Akintoye, 2000) identified several key issues including complexity of the project, scale and scope of construction, market condition, method of construction, site constraints, client's financial position, buildability and location of the project. It is believed that these factors have a direct effect on the productivity levels on site and the overall performance of the project.

Additional research by Akintoye and Fitzgerald (2000) identified that the most significant factors resulting in inaccurate estimates included insufficient time for tender preparation; poor tender documentation; insufficient analysis of the documentation by the estimating team; low level of involvement from the site team that will be responsible for construction; poor communication between the estimating and construction teams and lack of review of cost estimates by company management. Tendering for work in an area which contractors have little knowledge is also a significant reason leading to inaccurate estimating (Carr, 1989). Thus, when tendering for work overseas, UK contractors sometimes form joint ventures with home-based contractors.

Smith (1995) identified three main factors which could lead to inaccurate estimates: first, inappropriate assessment of risk; second, inappropriate contract strategies; third, human characteristics of the individual estimator.

Despite these challenges, however, there can be scope for innovation when tendering. It is not unknown, on major civil engineering projects, for the award to be made to the contractor who devises a more economic design than the one proposed by the engineer for part or even the whole of the works. Furthermore, contractors have been awarded contracts even though their tenders were not the lowest; this may be due to a highly original method statement and design solution for the temporary works. On major international infrastructure projects it is not unknown for the major *big six* Japanese contractors to offer the client a deferred payment scheme. Japanese firms will often bid low and provide an excellent service in the hope of establishing a new, long-term, loyal customer.

Detailed investigation and pre-planning is essential as considerable site overhead costs can be saved by early completion of the project including salaries of site staff, accommodation, site services, standing construction equipment etc.

However, before any estimates can be submitted the first step for contractors is to get onto the tender lists. This could be done on an ad hoc basis or preferably in accordance with a sort of longer-term strategic marketing plan that is usually of a five-year duration. The five-year plan, which should be based on an analysis of the past and consideration of the future trends within the market, should be re-examined on an annual basis and modified accordingly.

Panel 6.1 Case study: Cessnock Dock Railway, Glasgow

Winning contracts through innovation is not a new concept. In 1893 the original Robert McAlpine won a contract for the Cessnock Dock Railway in Glasgow with a tender far below any other. Everyone agreed that it would be impossible for him to do the work for the tender as he had no means of transporting the substantial amount of excavated material from the project to the sea where it was required to be dumped.

Excavations on the job were made in plastic clay, in shale and in sand. A brick-making plant was constructed and in the event every brick used in the project was made from the excavated material and millions of bricks were sold. Roads were constructed to the sandpits and McAlpine not only had the sand hauled away for him but was paid a fair price.

Source: Saxon Childs, 1925.

6.2   Stage 1 – decision to tender

The first stage in the tendering process is the decision to tender. As soon as the tender documents are received the estimator should quickly skim through the documents in order to establish the following:

1. Amount and type of work involved in project and whether the company has any competitive advantage;
2. The approximate value of the project together with a review of the major resources required, particularly construction equipment, staff, key subcontractors and suppliers;
3. The programme requirements, that is, completion dates, sectional completion and critical milestone dates – will these require excessive overtime?
4. The form of contract, specification, method of measurement and if any amendments have been made to standard documents?
5. The time and recourses required for preparation of the tender;
6. Whether any contractor's design is required, and whether the main contractor is required to accept liability for any subcontractors' design. The contractor should be provided with a design brief and a performance specification clearly identifying their design responsibility;
7. Possible alternative methods of construction (where the contractor's expertise leads them to consider that a more economical design could be used, for example precast piles in lieu of cast in situ);
8. Design for temporary works (including support structures/cofferdams/temporary bridges/river diversions/special shuttering/scaffolding/groundwater control systems etc.);
9. Whether the risks are acceptable – this could include the following: weather conditions/flooding/suitability of materials – particularly filling material from quarries/reliability of sub-contractors/non-recoverable costs, for example excesses on insurance claims/outputs allowed such as productivity/cost increases/terms and conditions in contract/ability to meet specification for price allowed/availability of labour/plant) and whether the tender is of particular interest.
10. Contract requirements for performance bonds, warranties and parent company guarantees;
11. Funding requirements for the project. It will be necessary to plot income against expenditure using the programme of works and the bill of quantities (BofQ)/priced-activity schedule – payments in arrears and retentions, both from the employer and to suppliers and subcontractors must be considered.

Upon completion of the review the estimator should complete a pre-tender data sheet, grade the tender based on the interest to the company and recommend whether or not to tender. If a contractor decides not to tender, the documents should be returned to the employer; however, in practice this rarely occurs.

The technical process of predicting the net cost of the works is carried out by a team comprising the estimator, planning engineer, materials estimator, estimating technician, together with possible contributions from temporary works designers and an experienced construction manager if the work is of a specialist or complex nature. At the end of the process the team will produce the *cost estimate*. The *estimate* is the prediction of the cost to the contractor. The *tender* is the price submitted by the contractor to the employer.

6.3   Stage 2 – determining the basis of the tender

During this stage the estimator, prior to the preparation of the cost estimate, will disseminate and assemble the key information and generally become familiar with the documents. Unlike pricing a BofQ in the building sector, a civil engineering BofQ/priced activities can be priced only when read in conjunction with the engineer's drawings and the specification. Projects carried out for water authorities, railway or road transport authorities are normally in accordance with standard sector-specific requirements.

Enquiries will be sent to subcontractors and major materials suppliers, the latter often based on the quantities calculated by the contractor's quantity surveyor from drawings. The contractor should also check that the major quantities in the BofQ are correct; if any are found to be incorrect this factor will be considered later at the commercial appreciation stage.

However, the most important part of this stage is for the team to determine the *construction method* and sequence upon which the tender is based, together with an *outline programme of the works*, two items that are inseparable.

The construction method will often be dependent on the design of the temporary works necessary to enable the permanent works to be constructed. Temporary works are normally designed in-house by the contractor, but in case of scaffolding and falsework, design may be supplied by specialist contractors.

Temporary works may have considerable time and cost implications, and can include cofferdams and temporary bridges on riverworks, temporary piling and jetties on marine projects, overhead gantries on elevated motorways, dewatering systems and grout curtains on deep basements etc.

The contractor may further be required to design part of the works to meet a performance specification, for example, concrete specified to strength or piling to a load-carrying capacity; this design would often be undertaken by specialist suppliers or subcontractors.

A further involvement of the contractor's design department may be in identifying more economic alternative design solutions for sections of the permanent works. This is often done in the hope of sharing the saving involved, which could be considerable.

The programme could be in the form of a bar chart or in the case of major works based on a network showing the critical path produced utilizing computer software. The programme will be used by the successful contractor as a control document for monitoring progress and calculating the effects of delay and disruption to the flow of the works.

The programme is particularly important as 15–40% of the cost of civil engineering work is time related and many items such as site overheads are computed directly from it. Furthermore, as most contractors will be bidding using the same quotations from subcontractors, hirers of construction equipment and materials suppliers, obtaining a saving in time is one of the few ways in which the contractor can show a substantial saving to the project cost.

During this stage the estimator will need to identify any inherent restrictions (e.g. delivery of bulk materials by rail or water) and any items on long delivery (e.g. special equipment). They will further need to consider alternative methods of construction, sequence of construction and the level of utilization of resources.

In accordance with the requirements of the standard conditions of contract the contractor is deemed to have inspected the site and examined its surroundings. The contractor may also visit the engineer's office and the local authority in order to examine core samples, location of existing services, traffic requirements and any other relevant information available. If the contractor is to be responsible for the design of a significant part of the works then the contractor may need to carry out further site investigations. However, it is the employer who should identify how this can be done most efficiently.

Following the site visit a standard, comprehensive pro-forma checklist will normally be prepared listing such items as access to site, site security, provision of services, soil and ground-water information, nature of excavation fill and disposal, nearest tipping facilities, availability of labour, construction equipment and materials, site organization and layout, land purchase for borrow pits etc.

Method statement
Contract: New Reservoir, Bryn Gwynant   Sheet No.:
Tender No.:   Prepared by:
                                                             Date:

No.	Operation	Quantity	Method	Sequence of operations	Plant and labour	Output	Duration
1.	Strip topsoil	7,500 m³	Excavate using loading shovel and transport to temporary tip using dump trucks		Loading shovel and three dump trucks	50 m³/hr	15 days
2.	Drainage To dam	300 m	Excavate using back actor, load, transport to temporary tip using dump trucks	Excavate, load trench support, lay pipes, backfill	Backactor +three dump trucks	Based on 3 gangs 15 m per day	20 days

6.1 Typical method statement.

The method statement (see Fig. 6.1) is a key document in the preparation of the tender and should consider the site-visit report, the geotechnical report, the sequence and methods for the main operations of work, subcontracted work, bulk quantities, schedule of labour and construction equipment and any temporary works required.

The Construction (Design and Management) Regulations 2007 (CDM2007) also require the principal contractor before the start of the construction phase to prepare a construction-phase plan which is sufficient to ensure that the construction phase is planned, managed and monitored in a way which enables the construction work to be started as far as reasonably practicable without risk to health and safety. The CDM2007 Regulations also require the principal contractor to ensure that the construction-phase plan identifies the risks to health and safety arising from the construction works and includes suitable and sufficient measures to address such risks.

Stage 2 can be summarized as follows:

- *Examine key documents*: drawings/specification/BofQ (works information, site information, contract data);
- Send enquiries to major subcontractors and materials suppliers;
- Check major quantities in the BofQ;
- Determine the method of construction and the outline programme of the works;
- Examine more economic alternative designs, design temporary works and any necessary permanent works;
- Identify inherent restrictions for example access to site/transport;
- Visit engineer's office – examine core samples;
- Visit site – compile site-visit report.

6.4   Stage 3 – preparation of cost estimate

During this stage the estimator will assemble information on the net cost of the works including calculating the following: the current rates for labour materials and construction equipment, the unit or activity rates, the preliminaries or general items and finally the summaries.

Current rates for labour, materials and construction equipment

The rates for labour will be the 'all-in' rates, that is, the contractor's total cost per hour of employing the different categories of labour. These hourly rates are calculated based on the basic rates as the national working rule agreement with the defined allowances for special skills together with bonus payments, holiday pay, CITB levy, employers' insurance etc. An example of the detailed build-up of the all-in labour rate is shown in the latest copy of E&FN Spon's *Price Book for Civil Engineering Works*.

The rates for material should cover for transport to site, offloading/storage, unavoidable double handling and waste. Prices for bulk materials must be scrutinized in order to ensure that they meet the specification and testing/sampling requirements; delivery must also meet the demands of the programme.

The construction equipment rates should cover for transport to site, erection/dismantling, operators, maintenance and fuel. Major static items of construction equipment such as tower cranes are normally priced separately in the general items or method-related charges section whilst other items are often included in the individual rates.

Unit rates for each item in the BofQ/activity schedule

The three main estimating techniques used by contractors when pricing major construction works are detailed below.

Unit-rate estimating

Unit-rate estimating, which is the standard procedure in the sector, involves pricing individual rates in the BofQ which has been prepared in accordance with a method of measurement for example SMM7. The unit rates are calculated using one of the following methods:

- Historical rates based on productivity data from similar projects;
- Historical rates based on data in standard price books for example E&FN Spon's, Wessex, Laxtons, Hutchins *UK Building Costs Blackbook*;
- *Built-up* rates from an analysis of labour, materials and construction equipment for each item and costed at current rates.

There are several possible disadvantages of using the unit-rate method for estimating major works. The system does not demand an examination of the programme or the method statement and does not encourage an analysis of the real costs and major costs risks in undertaking the work. Furthermore, the precision and level of detail in pricing each item can give a false sense of confidence in the resulting estimate.

Generally, it is not recommended that the data from standard price books are used in the estimating of major civil engineering works, either at tender or when variations are required. The reason for this is due to the possible differences in ground conditions, method statements, temporary works, availability of construction equipment, location of the project and the time of year in which the work is executed etc. Each project should be considered on its own merits and the cost estimate based on first principles using the operational method.

Operational estimating

Operational estimating, which is the recommended method for estimating civil engineering works, requires the estimator to build up the cost of the operation based on first principles, that is, the total cost of the construction equipment, labour and permanent/temporary materials. This method of estimating links well with the planning process as it embraces the total anticipated time that the construction equipment and labour gang are involved in the operation including all idle time.

If a BofQ approach is used, the total cost of the operation is then divided by the quantity in the BofQ to arrive at an appropriate rate. A significant advantage of this approach is that it provides a complete integration between the estimate and the programme which in turn enables a project cash flow to be produced. Furthermore, the pricing of operations, or activities, is obviously compatible with the NEC ECC *Priced Activities* approach. The process involves the following:

- Compiling a method statement, showing sequence, timing, resources required;
- Refining the method statement to show an *earliest completion* programme with no limit on resources;
- Adjusting the programme by *smoothing* or *levelling* the resources in order to produce the most economic programme to meet the time constraints;
- *Applying current unit costs*: fixed, quantity proportional and time related.

Establishing realistic productivity levels for labour and construction equipment on major operations can prove difficult, particularly on overseas work. However, the operational estimating approach enables the estimating team to better appreciate the major risks and uncertainties in the work.

Man-hours estimating

Man-hours estimating is most suitable for work which has significant labour content and/or for which extensive reliable productivity data exists for the different trades/specialisms involved. Typical applications include the following:

- Design work and drawing production, both engineering and architectural;
- Installation of process plants and offshore modules.

This method of estimating is frequently used by the major mechanical and electrical contractors as well as by the large American contractors, for example Bechtel. It should be used in conjunction with a construction programme/schedule in order to highlight any restrictions, for example availability of heavy-lifting equipment, which may affect labour hours expended in fabrication yards or on site.

AN EXAMPLE OF OPERATIONAL METHOD OF ESTIMATING

The question relates to the construction of a reinforced concrete basement (size 50 m × 30 m × 10 m deep) built below ground on a green field site.
   The contractor's estimator is required to calculate an appropriate BofQ rate.
   E326 Excavation for foundations, material other than topsoil,

rock or artificial hard material maximum depth 5–10 m   15,000 m³

Approach: Consider two alternative construction methods.

- Method A – open cut with battered sides (assume total volume of excavation equals 2.5 × net volume) – the open-cut method will require additional working space to allow for erect and strip shutter to the outer face;
- Method B – steel cofferdam built around net perimeter of basement.

Assume the following net costs (based on quotations from subcontractors):

- Excavation open cut – £10 per m³;
- Disposal on site – £1 per m³;
- Bring back and fill – £2 per m³;
- Excavation restricted within cofferdam – £25 per m³;
- Sheet piling (assume 15 m deep) – £35 per m³;
- Mobilization/demobilization-piling rig – £5,000 each way;
- Extract cofferdam piling – £5,000;
- Site overheads – 10%, head office overheads and profit – 12%.

Solutions

There are two solutions; costs involved as described in method A (open cut) and costs involved in method B (steel cofferdam) (see Figs 6.2 and 6.3); both are discussed in detail in the following sections.

Costs of method A (open cut)

6.2 Method A (open cut).

Excavating in open cut 15,000 × 2.5 m³ = 37,500 m³ @ £10/m³	£375,000
Disposal on site 37,500–15,000 = 22,500 m³ @ £1/m³	£22,500
Bring back and fill 22,500 m³ @ £2/m³	£45,000
Total net cost	£442,500

It is up to the contractor to select the most economic method of working. The additional excavation required is dependent on the nature of the ground and the natural slope of inclination – generally the harder the material the steeper the slope.

Costs of method B (steel cofferdam)

6.3 Method B (steel cofferdam).

Sheet piling – mobilization/demob 2 × £5,000	£10,000
Sheet piling 160 × 15 = 2,400 m² @ £35/m²	£84,000
Excavate within cofferdam 15,000 m³ @ £25/m³	£375,000
Extract cofferdam	£5,000
Total net cost	£474,000

Thus, based on the above, the estimator would choose the open-cut method.

Net cost of open-cut method	£442,500
+ 10% site overheads	£44,250
	£486,750
+ 12% head office overheads and profit	£58,410
	£545,160

So rate to be included in the BofQ should be £545,160/15,000 m^3 = £36.34/m^3

Subcontractors

The management of subcontractors can make or break a contract as typically subcontractors comprise over 75% of the total work executed. Few contractors have the necessary continuity in projects to justify purchasing specialist plant or have the expertise necessary to execute all the work, for example, specialist piling or diaphragm walling, hence the use of subcontractors. Indeed, over the years the role of the main contractor has shifted from a traditional works contractor towards the management of works packages.

However, it is still a prudent policy for main contractors to price as much of the basic sub-contract work as if it were being carried out by their own resources, this should ensure that the work is adequately priced and reduce the risk to the main contractor.

The construction industry has traditionally had an uneasy relationship with its subcontractors often going back to subcontractors in order to gain price reduction or increased discounts after tender award. Furthermore, main contractors choose to dump as much risk as possible onto subcontractors through the use of penal clauses in the subcontract documents. However, the more enlightened clients are now adopting a partnering approach with integrated project teams and fair conditions of contract, for example, the NEC ECC subcontract form.

Enlightened contractors could conduct a buildability review with the specialist designers in order to identify more economic/practical/safe methods of working. The developed improvements might or might not be declared at tender depending on the particular terms of the contract. For example, under a target-cost contract it may be advisable to leave the identification of these savings until the construction stage.

General items (preliminaries)

The general items or preliminaries represent the cost of operating the site and will need to be calculated separately. These items can be included on a time-related or a fixed-price basis. Typical general items on a major civil engineering project might include the following:

- Site staff, including project manager, agents, engineers, foremen, quantity surveyors, office manager, store keeper, clerks, secretarial;
- Head office staff allocated to project, for example, designers; health and safety;
- Company cars;
- Site offices, mess huts, toilets, running costs;

- Transport for construction equipment;
- General site labour;
- Services connections and running costs;
- Haul roads;
- Temporary fencing;
- Construction equipment purchases including personnel carriers, land rovers, compressors, pumps, cranes, miscellaneous;
- Scaffolding and hoists;
- Access for subcontractors;
- Small tools;
- Plant consumables including fuel and fuel distribution;
- Contract works insurances;
- Setting up compounds;
- Security;
- Signboards;
- Road-cleaning facilities;
- Computing equipment;
- Office stationery etc.

If the contractor is successful and is awarded the contract then detailed records will need to be kept for these preliminaries items. If any extension of time claims are made these items will form the basis of the *site overheads* component within a prolongation claim.

Preparation of summaries, tender summary, analysis sheets, special conditions

Self-explanatory but basically the bringing together of the estimate. Stage 3 (preparation of the cost estimate) can be summarized as follows:

Calculate current cost rates for labour/materials/plant

- Calculate unit rates for each item using one of the following methods:
  - a.   Operational method (based on method statement and construction programme – consider total resources, that is, teams of operatives and equipment);
  - b.   Unit rates (normal approach in the building sector);
  - c.   Man-hours estimating (used by US contractors on Petrochemical projects – based on work study feedback).

- Analyse and check subcontractors' quotations;
- Price preliminaries;
- Prepare summary sheets.

6.5   Stage 4 – commercial appreciation

Tender committee meeting – part 1

Following the production of the cost estimate a small management team, comprising the chief estimator and proposed contracts manager, will make a separate comprehensive evaluation of the estimate to ensure that the bid is both feasible and commercially competitive.

The first task of the senior management team at this tender committee meeting is to review the estimate taking into account the construction method and programme, the technical and commercial risks, the contract cash flow and finance, the potential for use of own construction equipment, the competition, the economic climate and the commercial opportunities.

On contracts involving major earthworks the risk can be considerable, particularly in connection with borrow pits and quarries, as the material to be extracted may subsequently be rejected as unsuitable by the engineer, or the local authority may refuse a planning application for the extraction.

Weather conditions can also be influential with continuous wet weather likely to cause a prolonged shutdown of all major earthmoving operations, the costs of which may not be recoverable under the contract.

Other risks include reliability of subcontractors – failure to perform or bankruptcy is at the contractor's risk; increases for inflation, for example steel and fuel; availability of suitable labour – specialist labour may need to be brought in from elsewhere within the UK or from overseas; terms and conditions in suppliers or construction equipment contracts; estimator's productivity allowance; premiums on insurances.

The team will also consider the commercial opportunities, particularly the method of billing and whether the major quantities are under or over measured or any items omitted entirely, any differences between the specification, drawings or B of Q, the lack of drawings or poor design and the contractor's alternatives. This stage can be summarized as follows.

Review the following:

- Method statement;
- Programme;
- Technical, design and commercial risks (NEC3 has introduced the concept of a risk register which allows contractors to identify which items they have/have not allowed for);
- Cash flow and finance;
- Use of own construction equipment;
- Competition;
- Commercial opportunities;
- Economic climate.

6.6   Stage 5 – conversion of estimate to tender

Tender committee meeting – part 2

The second task of the senior management team at the tender committee meeting is to convert the estimate into the tender bid. The following items are considered and agreed upon:

- The financial adjustment to be made following the commercial review;
- The allowances for discounts on subcontractors and suppliers;
- Late quotations, these could be included as an *adjustment* item at the end of the tender;
- The contribution for head office overheads – usually between 4% and 8%;
- Profit, normally on what the market can stand;
- Qualifications to the bid, if any.

At the conclusion of the meeting the estimator will be required to convert the cost estimate to the tender bid.

The difference between the 'prime cost' (the cost of doing the work in the field including all operatives below foreman level (labour/materials/plant/subcontractors) and the 'final tender' is called *the spread*, which can amount to 25% of the tender. This amount can be allocated to the tender in a number of ways:

- Evenly on all rates;
- Differentially, for example, front-end loading on early items;
- As a lump sum in the preliminaries.

6.7   Stage 6 – submission of tender

Finally the tender should be submitted to the client in the form specified in the invitation letter, arriving at the correct address at the right time. The contractor should keep all copies of the tender documents marking drawings 'used for tender'.

6.8   Conclusion

Establishing the cost of the estimate and the price of a traditional, or design and build project, in the civil engineering sector presents a massive challenge to the contractor.

In the UK, fewer clients in the civil engineering and infrastructure sector are using the standard ICE *Conditions of Contract* with tenders based on detailed BofQs. Instead, many clients are now choosing to use the NEC engineering and construction contract's priced-activities route. This approach complements the operational method of estimating but often requires the contractor to calculate their own quantities.

Using the data generated from the operational method of estimating, and the information received from specialist subcontractors, the estimator should be able to prepare a detailed method statement. This key document identifies the activities and the methods of working, the labour gangs and construction equipment, the temporary works, and critically, the production rates (concrete output per hour or day) enabling the calculation of total days per activity. The method statement thus provides the key data which forms the basis for the construction programme (or schedule).

All estimates will require the estimator to assess the risks for the project, which on civil engineering projects can be considerable, and forecast future increases in raw materials, if it is a fixed-price tender. Traditionally, contractors have sought to shift these risks onto subcontractors through the use of harsh conditions of subcontracts.

Furthermore, in the past, many contractors were selected on the basis of the lowest bids. These low bids might have been developed based on innovative working methods or alternative designs but might also have been a result of the contractor's misunderstanding or underestimating of risks in the project. This traditional approach often resulted in an adversarial position being taken by both the client and the contractor.

In order to improve the process, we have seen in Chapter 4 how some clients are now selecting contractors on the basis of a Quality/Cost model typically (60/40%). For one-off projects, some clients are selecting contractors based on a design and build approach under a partnering arrangement. Other clients, with a long-term building programme, are utilizing framework agreements with payments made based on target costs under an open-book accountancy approach. So the world is rapidly changing, but the estimating and tendering approach described in this chapter will still have relevance for many projects, including PFI schemes, both in the UK and overseas.

6.9   Questions

1.   Why would a promoter wish to make amendments to the standard contract documents?
2.   Why would an estimator go to the trouble of checking the major quantities?

3. Labour costs are estimated based on an *all-in* rate. What is meant by an *all-in* rate and how is it calculated?

4. How would the contractor's estimator calculate a due allowance for waste on materials and how accurate is this likely to be?

5. What are the main elements of cost, which must be taken into account a tender for civil engineering works and how would the estimator assess each item?

6. A tender for a project requires considerable deep excavation for which you propose allowing the use of temporary sheet piling. However, you are aware that battered side slopes may be a suitable alternative at a similar cost. How would you insert the costs in the tender BofQ/activity schedule?

7. Describe the steps you would take as a contractor to ensure that before submitting a tender you had obtained for yourself all the necessary information as to site conditions, risks, contingencies and all other circumstances influencing or affecting your tender, as required under the NEC ECC contract.

Bibliography

Akintoye, A. (2000) 'Analysis of factors influencing project cost estimating practice', *Construction Management and Economics*, vol. 18, no. 1, January, pp. 77–89

Akintoye, A. and Fitzgerald, E. (2000) 'A survey of current cost estimating practices in the UK', *Construction Management and Economics*, vol. 18, no. 2, January, pp. 161–172

Al-Harbi, K.M., Johnson, D.W. and Fayadh, H. (1994) 'Building construction detailed estimating practices in Saudi Arabia', *Journal of Construction Engineering and Management*, vol. 120, no. 4, pp. 774–784

Carr, R.I. (1989) 'Cost estimating principles', *Journal of Construction Engineering and Management*, vol. 115, no. 4, pp. 545–551

The Chartered Institute of Building (CIOB) (1997) *Code of Estimating Practice*, 6th edition, Longman

Childers, J. Saxon (1925) *Robert McAlpine. A Biography*, University Press Oxford

Civil Engineering and Highway Works Price Book (2007) latest edition, E&FN Spon

McCaffer, R. and Baldwin, A. (1991) *Estimating and Tendering for Civil Engineering Works*, BSP Professional

Seeley, I.H. (1993) *Civil Engineering Contract Administration and Control,* Second Edition, Macmillan Press

Smith, N.J. (1995) *Project Cost Estimating*, Thomas Telford

Part III
Key tools and techniques

7 Value management

7.1 Introduction

The value management process (see Fig. 7.1) originated during World War II within the General Electric Company (GEC) in the USA. GEC was faced with an increase in demand but had a shortage of key materials. Larry Miles of GEC, instead of asking 'how can we find alternative materials', asked 'what function does this component perform and how else can we perform that function?' This innovative approach led the company to use substituted materials for many of its products. They found, surprisingly, that the cost of the product was often reduced but the product improved; care and attention to function provided *better value for money*.

A spin-off of this approach was the elimination of cost, which did not contribute to performance – this was known as value analysis. Over the next ten years this was further developed by GEC and became known as value engineering (VE). Value management (VM) developed from VE and is now a requirement on many public and private projects in the USA and Australia.

It was only in the late 1980s that VM began to be used in the UK. The author first came across the use of VM on the £50m International Convention Centre in Birmingham; this four-year project was completed in 1991. Two years into the construction period an American VM consultant was engaged to execute VE exercises. However, by that time it became too late to effect any meaningful changes. In reality, there are various triggers for a VM exercise, which are usually workshop based, for example new legislation, new opportunities for a commercial product, solution of a social problem or simply overspend on the budget.

In the UK, the public sector has been slow to take up VM; however, with the introduction of concepts such as 'Best Value' and 'Prime Contracting' there has been an uptake in interest. However, the ultimate challenge is to integrate risk management and VM into a single framework that evolves throughout the life of the project.

7.2 What is value management?

Value management (VM) is the wider term used in the UK to describe the overall structured, team-based approach to a construction project. It addresses the value process during the concept, definition, implementation and operation phases of a project. It encompasses a set of systematic and logical procedures and techniques to enhance project value throughout the life of the facility.

VM embraces the whole-value process and includes value planning (VP), VE and value reviewing (VR).

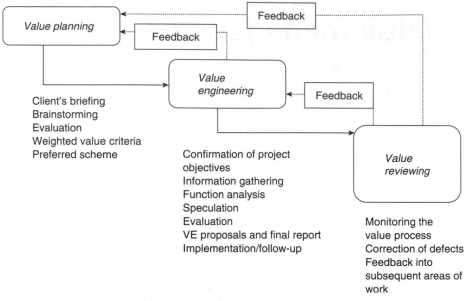

Client's briefing
Brainstorming
Evaluation
Weighted value criteria
Preferred scheme

Confirmation of project
objectives
Information gathering
Function analysis
Speculation
Evaluation
VE proposals and final report
Implementation/follow-up

Monitoring the
value process
Correction of defects
Feedback into
subsequent areas of
work

7.1 The value management process (source: ICE, 1996).

The basic steps are as follows:

1. To determine the functional requirements of the project or any of its constituent parts;
2. To identify the alternatives;
3. To examine the cost and value of each alternative to enable the *best value selection.*

Terminology

Confusion has arisen in the definitions of value depending on geographic location; VM/VE, VP, value auditing are often interchangeable. The following definitions are preferred:

- **Value:** value is the level of importance that is placed upon a function, item or solution.
- **VE:** value engineering is the process of analysing the functional benefits a client requires from the whole or parts of the design.
- **VP:** value planning is applied during the development of the brief to ensure that value is planned into the project from the beginning.
- **VM:** VM is a systematic and creative procedure operating on the relevant aspects of the value process through the life of the project or facility.
- **Function:** a mode of action or activity by which a thing fulfils its purpose. Understanding the concept of function is important as this can provide the catalyst to introducing innovative solutions. For example, consider the function of an internal wall, it can: separate space, secure space, maintain privacy, support heating systems, support fittings and fixtures, transfer load, reduce noise etc. If we merely required to separate floor space we could use a row of potted plants or different floor material.

Figure 7.2 shows the introduction of the parties into the project cycle under the traditional procurement route; this highlights the importance of involving all the key parties early in the process under a partnering agreement or better still a long-term alliance. Studies at the early stages of a project are much more effective and of shorter duration than those conducted later on. This is because the opportunities for making changes reduce as the project progresses, and

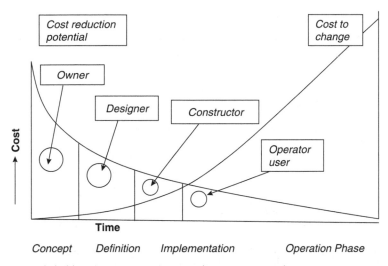

7.2 Stakeholders' impact on project cost (source: ICE, 1996).

the cost of making such changes increases. Indeed, once the concept has been frozen, about 80% of the total cost has been committed – even though no design exists.

All client bodies operate a capital-approval process that calls for certain criteria to be met before passing from one stage to the next – known as approval gateways. Each approval gateway presents a natural opportunity to conduct a VM study to verify that the scheme, as it has evolved so far, represents optimum value to the client. It is unusual to conduct a formal study at all of these gateways – usually two, or at most three, are sufficient.

The first stage in any project is to establish that a project is the most appropriate way in which to deliver the benefits which are required and then consider the following questions. Is it likely to be viable? Do the conditions exist to enable the project to stand a chance of success? Is it affordable? Answering these sorts of questions is the main purpose of Gateway 0 in the OGC's Gateway Review Process. VM can play a significant contribution at this stage. Table 7.1 identifies the key questions which should be asked at each stage of the VM study throughout the project cycle.

7.3 Value planning (VP)

Value planning (VP) is applied during the concept phase of a project. VP is used during the development of the project brief to ensure that value is planned into the whole project from its inception. Several outline designs are assessed to select a preferred option which best meets the functional and other requirements.

At this stage the value criteria are identified and concept proposals are put forward to satisfy the client's needs and wants. The needs are those items which are fundamentally necessary for the operation of the project, whereas the wants are items which the client would like to have but are not essential. Best value is provided by delivering a solution which delivers all the needs and as many of the wants as possible within the permitted budget.

In order to achieve maximum benefit from the effort applied by the individuals, it is common practice to apply the principle of the Pareto rule (80% of the value lie in 20% of the items).

Table 7.1 Typical questions to be asked at each stage of the value management (VM) study on a new urban highway project.

Stage in project	Questions to be asked	Comments
Concept	What is the problem?	Road congestion, lack of decent infrastructure
	Is this the right project to deliver the benefits?	Yes
	Does this meet our business criteria?	Part of government/local transport plan
Feasibility	Which is the best option?	Route B
	Does this solution satisfy our need?	Yes, least demolition of existing housing
Design	Does the solution fulfil all requirements?	Best compromise solution
	Is it good value for money?	As good as possible
	Can it be built?	Yes
Construct	Are the components cost-effective?	Detailed value engineering (VE) exercise required on road surface, bridge and tunnel construction
Use	Did we achieve the expected benefits?	Improved transport links now helping to regenerate city
	Are there lessons to be learnt for our next project?	Develop-design-build partnering approach; always expect the unexpected!

Source: Adapted based on Dallas, M., 1998.

7.4   Metropolis United's new football stadium

Second division Metropolis United is keen to move from their cramped town centre stadium to a greenfield site on the edge of town. The directors of the club realize that they require a 30,000 all-seater stadium if they are to compete in the top division.

The client's project manager suggests that a VM exercise would enable the directors and the council officers representing the local council who are partly funding the project to identify the priorities for the club. Unfortunately, there is little involvement of other key stakeholders – the fans, the manager and the players! The mechanism for incorporating best value into the project design is through the use of VP workshops. The first step of the workshop is to gather information concerning the project – generally through a briefing with the client.

In the first stage of VP, at the Concept Stage, a *value hierarchy* (see Fig. 7.3) is developed. It aims to establish a shared perception of the design objectives and attributes.

In the second stage of VP a *value tree* is drawn up (see Fig. 7.4) based on a simplified hierarchy. Although capital cost is one of the attributes used to evaluate design options, it is preferable to omit it from the value hierarchy and deal with it separately at the end of the analysis.

The logic of the diagram emanates from the how–why approach. In essence, by providing all the criteria on the right-hand side of the diagram, one will have provided all primary requirements on the left-hand side of the diagram. These criteria can then be weighted according to their degree of importance to the client.

In the second stage of VP, the project solutions would have been proposed which met these criteria; usually this exercise takes the form of a brainstorming session where creative thinking and synergy between experienced participants lead to effective solutions to meet the value criteria.

After some considerable deliberation, the client representatives compile a weighted value tree (with the highest attribute scoring 50 and the lowest scoring 10). Naturally the directors of the club, who are providing their own financial support to keep the club in existence, consider

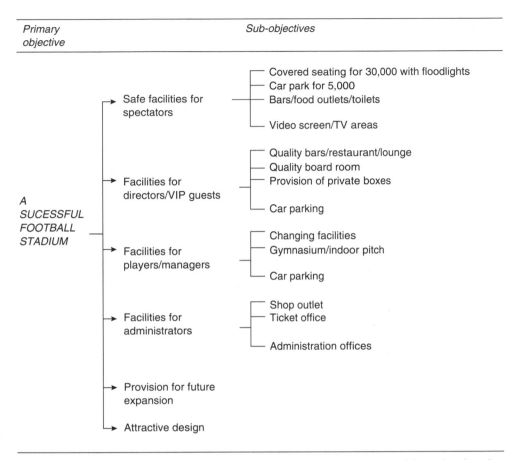

Primary objective	Sub-objectives

A SUCESSFUL FOOTBALL STADIUM

- Safe facilities for spectators
  - Covered seating for 30,000 with floodlights
  - Car park for 5,000
  - Bars/food outlets/toilets
  - Video screen/TV areas
- Facilities for directors/VIP guests
  - Quality bars/restaurant/lounge
  - Quality board room
  - Provision of private boxes
  - Car parking
- Facilities for players/managers
  - Changing facilities
  - Gymnasium/indoor pitch
  - Car parking
- Facilities for administrators
  - Shop outlet
  - Ticket office
  - Administration offices
- Provision for future expansion
- Attractive design

7.3 Value hierarchy at concept stage for new football stadium (note: Developed by author based on CIRIA, 1996).

that the facilities for directors and visiting VIPs are paramount; these attributes are therefore scored with a high 40 or 50. Attractive design is considered low priority for these self-made businessmen and is marked at 20.

The sum of the initial scores is 540. Each of the weightings of the six secondary objectives is then calculated. Thus a *safe facilities for spectators* equates to 130 (40 + 30 + 30 + 30) divided by the total score of 540 equals a weighting of 0.24% or 24% of the total.

In the next stage, the possible design solutions submitted by the design and build (D&B) contractors would be evaluated and ranked. The allocation of importance weights to the value hierarchy forms the basis for the next stage of the second workshop, which is deciding which of the available designs provide the best value. Evaluation involves assessing each option against each of the identified attributes, and this is best done in the form of a *decision matrix* (see Fig. 7.5). Finally, the team would make recommendations to the client.

Design option B shows the highest score with 54.0, with the highest score for *facilities for directors* but the lowest score for *safe facilities for spectators*. Design C shows a marginally lower overall score than B but with improved *spectator facilities* and a lower score for the *directors' facilities*.

This exercise is typical of the possible dilemmas facing clients and their advisors. This is not a science, but more an art. It is extremely difficult to score each of the sub-objectives and choosing

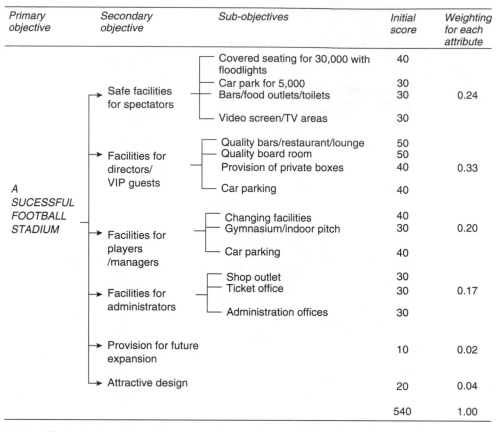

7.4 Simplified value tree at feasibility stage with importance weightings (note: Developed by author based on CIRIA, 1996).

Attributes	Safe facilities for spectator	Facilities for directors/VIP guests	Facilities for players	Facilities for administrators	Provision for future expansion	Attractive design	Total
Weight of importance	0.24	0.33	0.20	0.17	0.02	0.04	
DESIGN OPTION A	70	20	10	20	40	60	
	16.8	6.6	2.0	3.4	0.80	2.4	32.0
DESIGN OPTION B	40	80	20	60	70	60	
	9.6	26.4	4.0	10.2	1.4	2.4	54.0
DESIGN OPTION C	50	50	80	40	10	40	
	12.0	16.5	16.0	6.8	0.2	1.6	53.10

7.5 Decision matrix (note: Shows the process of comparing the total scores of the various design options).

between one contractor's design and another again becomes subjective. In the event, design option B is chosen!

7.5   Value engineering (VE)

VE is applied during the definition stage and, as required, in the implementation phases of a project. VE investigates and analyses in order to identify the required function and then compares and selects from the various options to produce the owner's best value requirements.

During the VE phase any unnecessary cost is eliminated from the proposed design. This is usually undertaken in the VE workshop where a separate review team from that which developed the outline design reviews the work to date. Since this review is generally undertaken at approximately the 30% stage, there is still a good opportunity to adjust the design before it proceeds to the definitive and detailed design stage.

The basic premise of VE is that a certain amount of unnecessary cost is inherent in every design. It is usually only possible to eliminate this by identifying another option, which provides the same function at lesser cost.

Specific causes of unnecessary cost include the following:

1. Cost of unnecessary attributes (attributes which provide no useful function);
2. Cost of unnecessary specification (due to needlessly expensive materials/components);
3. Unnecessary cost of poor buildability (failure to consider construction implications during design);
4. Unnecessary life-cycle cost (failure to consider future operational costs);
5. Unnecessary opportunity cost (the cost of losing potential revenue).

The VE workshop follows the broad principles of the VP workshop. The information phase usually involves a debriefing from the original design team to the VE team, who then consider, in the functional analysis, the function of each part of the proposed works. In the speculation phase, they have a brainstorming session and consider alternative methods of providing the same function.

There follows an evaluation phase in which the proposed alternative solutions to providing the function are analysed to determine the viability of each one. Where a suitably viable alternative is possible at a significantly reduced cost, it is included in the proposal phase.

Panel 7.1 Value engineering techniques

The different ways of delivering a client's requirements offer further potential for adding value to a project. With engineering services, examples of innovation that can have a significant effect on the overall project outcome include the use of ground-water cooling or gas-fired air-handling units.

Innovative solutions such as this need to be adopted at the earliest possible stage of a project. When the value engineering approach is applied at a later stage, it is difficult to introduce radical changes. However, opportunities to add value still exist, such as the use of manufacturer's standard components rather than bespoke products.

Source: www.davislangdon.com (accessed August 2005, value engineering cost model) For a description of small-scale renewable energy systems see *Building*, 28 October 2005, pp. 54–57.

Panel 7.2 40-hour workshop

In the USA, the classic value engineering (VE) exercise is a 40-hour workshop attended by the value manager and an independent design team. The findings are reported to the client and project manager for further action/implementation.

While an independent design team has the advantages of providing a fresh and critical approach and an independent review, in the UK the disadvantages are generally believed to outweigh them. These include

- conflict with the existing design team;
- loss of time while the external team becomes familiar with the project;
- the additional cost of a second design team;
- delay and disruption to the design process during the review.

Also the external team may feel obliged to identify cost savings to justify their fee.

Source: CIRIA, 1996.

Ideally, every design decision should be subject to VE, but 80% of cost is often contained in 20% of the design decisions. On building projects, services in particular account for a very large percentage of the overall cost (28–40%). This element can be further broken down into mechanical services (17–28%), electrical services (6–13%) and lifts (0–3%). On road projects, the three highest-cost elements are typically earthworks (28–31%), structures (18–32%) and sub-base and surfacing (21–28%) (CIRIA, 1996).

7.6   Value reviewing (VR)

VR is applied at planned stages to check and record the effectiveness of the value process and its management.

The 'value manager' usually has a responsibility to review the value process throughout the project to ensure that the value identified in the VP and VE are actually provided within the executed works.

When to apply VM

Timing is of the essence. Fig.7.2 illustrates the substantial scope to reduce cost, and hence improve value, in the project definition and early design phases. This scope diminishes to point when the cost of change exceeds the saving.

Any construction project should only be commissioned following a careful analysis of need. Failure to carry out this analysis will cause problems at subsequent design and construction stages.

Many projects suffer from poor definition through lack of time and thought at the earliest stages. This is likely to result in cost and time overruns, claims, user dissatisfaction or excessive operating costs. VM can help to avoid these problems.

VP and VE are mainly applied in the concept and definition phases, and generally end when the design is complete and construction started. However, VE can be applied at the construction stage to address problems or opportunities which may arise.

At a later stage a tendering constructor may be expected to bring other value improvement ideas and techniques for consideration by the owner.

Finally, the project may run into practical, cost or time difficulties during construction, and here again solutions may be developed using VM.

7.7 Case studies

Case study 1: the Turner Centre in Margate

The Turner Centre in Margate, Kent, is an ambitious and futuristic art gallery. It forms one of the first steps in a much broader programme to help regenerate Margate. Numerous stakeholders groups are involved, including Kent County Council and Thanet District Council, that comprise the *client*, government departments and agencies, regulatory bodies, funding organizations and the eventual managers of the facility.

  VM processes were used early on in the project to enable key stakeholders to build understanding and consensus regarding the goals of the project, how it related to the overall regeneration programme and to define the actions necessary to enable it to be realized.

Source: Dallas, M., 2005.

Case study 2: VM session, Midlands Hospital

The VM session was the first session of its kind involving the wider project team, which included the 'Procure 21' supply-chain members and a number of clinicians from various affected departments.

  The purpose of the session was to review, reflect and to understand some of the key project issues as well as to determine spatial adjacencies and linkages required with the rest of the hospital. The session was led by an eminent VM practitioner.

  The first tool in the session, the client's value system, was used to obtain ordinal measurement in the form of ranking for the client's values. The client's value system, based on a matrix approach, demonstrated that the three most important aspects to the client group were flexibility, comfort and community accessibility. The client's value system concentrates on 'inside the building', that is, the focus is how the building functions rather than its external appearance.

  The service development group reviewed typical patient flows through the new facility. This detailed analysis helped to achieve a consensus on requirements and a degree of ownership of the final solution.

  The project team then used a brainstorming session, using 'post-it' notes on a working wall to identify strategic needs and wants and technical needs and wants. The strategic functions identified the strategic mission for the project and the main functions that it requires to deliver for the client. The strategic wants were considered as non-essential or 'nice to have'. The technical needs and wants are the functions that form the brief for the project. The Procure 21 team agreed to incorporate the technical needs into the project brief as well as any strategic issues that were to be dealt with, by looking for a technical solution.

  The VM exercise took the project team a step further in validating a conceptual model for the new facility. Once this was finalized the Procure 21 supply team members could draw up the 'project execution plan'.

Source: VM Session undertaken in October 2002 (private source).

Case study 3: office building

The sketch design for the outside walls of an office building indicated the use of precast concrete-cladding panels. An analysis of the design showed a total of 450 separate panels of different types.

After a workshop review, which included a cladding manufacturer, the total number of panels was reduced to 280 with only 21 different types to cover the same area. Although an increased cost arose due to the need for a larger crane to hoist the panels, the net saving for the reduced number of moulds and perimeter waterproofing represented 10% of the total cost of cladding. The cladding element of the project amounted to 25% of the project's cost, and as a result of the workshop review, the overall saving equated to 2.5% of the project's capital cost.

Source: Reported in HM Treasury, 1996.

Case study 4: office building

The detailed design of a bolt-on cladding system indicated an internal wall lining of insulation and painted plasterboard. Following a brainstorming session, the cladding manufacturer was asked to provide a price for incorporating the insulation and providing a metal-finished panel on the inner face of the building. The plasterboard and its finish would be omitted.

The net effect was to increase the cost of the project by £125,000. However, omitting the plasterboard and paint meant fewer 'wet' trades on the project, saved three weeks on the overall construction period and increased the net lettable floor area by 2%. The *value* of the finished building was increased in the order of £2 million for an additional outlay of £125,000.

Source: Reported in HM Treasury, 1996.

Case study 5: hotel leisure facility

In a hotel development, the architect had included the main hotel swimming pool and a children's paddling pool. The design team had assumed that the function of the children's pool was to allow the children to swim separately from the adults, thereby providing a more suitable facility for each.

In fact, the function of the pool was to keep much younger children safe while others swam and there was no real objection to competent child swimmers using the same pool as the adults.

As a result of correctly defining the function, the VM team was able to generate ideas for alternatives providing a safe environment for the children. The small pool was replaced with a spray at a tenth of the cost. When constructed, the spray proved to be a huge success.

Source: Reported in McGeorge and Palmer, 1997.

Case study 6: speculative office block, London

A developer proposed to construct a speculative office building in London with a net lettable area of 4,500 m² and at a cost of £5 million. Initial studies indicated that providing the necessary space within the site and cost constraints would be challenging.

The developer decided to use VE and employed a facilitator to work with the design team to find the most effective solution. The facilitator convened a workshop during which it became clear that the relationship between net lettable area and the size of the service cores was crucially important to the viability of the design. Although the designers had already evaluated this, the combined efforts of all the parties working together creatively in a facilitated workshop environment identified a number of improvements to the outline design proposals. Potential improvements were also identified in the proposed wall-cladding system, and these were subject to a more detailed study outside the workshop.

As a result of the workshop, the building cost 2% (£110,000) more than the original budget. However, the increase in benefits of some £3.4 million more than paid for this.

Source: Reported in the CIRIA, Special Publication 129.

Case study 7: Dudley Southern Bypass

In 1998, Kvaerner won the project in competition with an exceptionally low bid of £14.3 million; the contract was based on the ICE 5th with a 'partnering agreement'. After a joint evaluation of the risks, the Dudley Metropolitan Borough Council felt able to negotiate a target price which would still be below the second-lowest bidder and a target cost of £16.7 million was agreed. Dudley MBC agreed with Kvaerner that it would split 50/50 any 'pain' or 'gain' over or under that target price and Kvaerner would be paid an agreed maximum management fee of £900,000. The project was completed five months ahead of schedule within the target cost and the budget agreed with the DETR.

VE did identify savings. For instance, the original specification required the removal of 50,000 m³ of waste to be replaced with quarry material. Much of the material was contaminated, but by working together and involving the Environment Agency in developing solutions, they were able to reuse most of the material within the project. By the end of the project, they had only taken 1,500 m³ to tip, this prevented 25,000 lorry movements around Dudley.

Source: Reported in *Modernising Construction*, NAO, 2001.

Case study 8: the Scottish Parliament Building

Construction work began in July 1999. In September 1999, a VE exercise was implemented in order to reduce the construction cost by £25 million. The exercise identified several hundred recommendations, the vast majority of which could be dealt with by the project team. Some of the recommendations required a high-level decision from the Scottish Parliament Corporate Body (SPCB). Table 7.2 gives an indication of the issues considered by the SPCB.

Source: Reported in *The Holyrood Inquiry*, 2004.

Table 7.2 Examples of VM savings considered on the Scottish Parliament building.

Item	Potential saving	Decision of the SPCB	Actual saving
Reduce car parking provision from 129 to 50	£750,000–£1,500,000	Reduce to 65 spaces	£667,000
Rationalize bar/lounge/restaurant	In excess of £1 million	Maintain existing provision	—
Delete wash-hand basins in MSP rooms	£210,000	Agreed to be deleted	£209,160
Reduce standard of media accommodation fit out	£235,000	Agreed	£236,140

Note: Developed by author based on CIRIA, 1996.

Lord Fraser reports that the exercise failed miserably in achieving its stated goal of achieving the £25 million required. The workshop never identified achievable savings of the magnitude required. Likewise, when the client realized that these decisions would have a significant impact on the quality of the building they did not face up to the reality of the situation. Lord Fraser comments that 'to some extent the Value Engineering exercise could be interpreted as a knee jerk reaction to a budgetary crisis.'

Case study 9: refit project for Pizza Hut

In 1997, Pizza Hut was anticipating a programme of 25 refit projects (Pizza Delivery Units). At an estimated/budget value of £145,000 each, the programme value amounted to over £3.5 million. In order to review the projects before going on site, a series of three half-day VM workshops were convened comprising client representatives (area manager and property manager) and the consultants (designer, QS and services engineer). An experienced VM facilitator from outside the project team facilitated the workshops.

   The three workshops followed a traditional format of information exchange, functional analysis, brainstorming of alternative solutions, evaluation of alternatives, acceptance and implementation. As a result of nine hours of workshop and a similar amount of time outside the workshop a total of £14,000 per project was saved equivalent to £350,000 capital cost on the whole programme. In some areas standards were actually raised, and longer-term maintenance was reduced. A shorter contract period was also established and shorter delivery times for certain longer lead items.

The total cost of the VM exercise was estimated at less than £10,000, giving a return of 35:1 on the investment.

Source: Reported in *Value Management Fact Sheet* (www.constructingexcellence.org.uk [accessed 12 December 2006]).

7.8 Conclusion

The above case studies demonstrate that real benefits and cost savings can be secured by implementing a VM approach. Case study 1 identified that VM helped bring the key stakeholders together in order to build an understanding and consensus.

Case study 2 identified how the VM approach enabled an NHS client to determine special adjacencies and linkages and identify the client's value system as a basis for validating a conceptual model for a new hospital facility.

Case studies 3 and 4 showed how significant savings could be achieved by redesigning an alternative size and type of cladding. As shown in Case study 5, Dr Angela Palmer brilliantly illustrated the classical benefit of the VM approach by giving an example in which the project team questioned the fundamental purpose of a children's paddling pool.

Case study 6, the 'speculative office block', illustrated how the VM exercise showed that a little extra expenditure would result in a significant greater net lettable area.

Case study 7 demonstrated the benefit of VE within a partnering approach on the Dudley Southern Bypass which resulted in a considerable saving in the removal of excavation waste.

Case study 8 demonstrated the difficulty of achieving real savings through VE on the hugely controversial Scottish Parliament building.

Finally, the Pizza Hut case study (Case study 9) demonstrated how VM, when used by an enlightened client resulted in cost savings, improved standards and shorter delivery times on 25 refit projects.

7.9 Questions

1. Consider the function of an external wall in an office complex.
2. Give specific examples of potential unnecessary costs identified in items 1–5 in Section 7.5 on 'value engineering'.
3. The University of Metropolis is planning a new state-of-the-art teaching facility for the School of the Built Environment. As the client's chosen project manager, write a 500-word report to your client, explaining the key concepts and benefits of including a VM approach and state what will be required from the client and when.

Bibliography

Connaughton, J.N. and Green, D.G. (1996) 'Value Management in Construction: A Client's Guide', Special Report 129, Construction Industry Research and Information Association (CIRIA)

Dallas, M. (1998) 'The Use of Value Management in Capital Investment Projects', Session Guide Television Education Network Video: Quantity Surveying Focus, March

Dallas, M. (2005) 'Delivering Best Value on Regeneration and Housing Projects' (www.davislangdon.com/pdf/EME/Publications/RM_VM/Delivering%20Best%20Value.pdf [accessed December 2006])

Dallas, M.E. (1992) 'Value management – its relevance to managing construction projects', in *Architectural Management* (ed. P. Nicholson), E&FN Spon, pp. 235–246

Dell' Isola, A.J. (1982) *Value Engineering in the Construction Industry*, 3rd edition, Van Nostrand Reinhold

Green, S.D. (1992) 'A SMART Methodology for Value Management', Occasional Paper No. 53, CIOB

Green, S.D. and Popper, P.A. (1990) 'Value Engineering: The Search for Unnecessary Cost', Occasional Paper No. 39, Charted Institute of Building (CIOB)

HM Treasury (1996) *CUP Guidance Note No 54: Value Management*, HM Treasury

ICE (1996) *Creating Value in Engineering – Design and Practice Guide*, Thomas Telford

Kelly, J. and Male, S. (1993) *Value Management in Design and Construction: The Economic Management of Projects*, E&FN Spon

Kelly, J., Male, S. and Drummond, G. (2004) *Value Management of Construction Projects*, Blackwell Science

McGeorge, D., Palmer, A. and London, K. (2002) *Construction Management: New Directions*, 2nd edition, Blackwell Science

Useful websites

www.constructingexcellence.org.uk
www.IVM.org.uk

8   Risk management (RM)

8.1   Introduction

Max Abrahamson, the eminent construction lawyer, considered risk management (RM) 'the most delicate and dangerous subject I could find' (Abrahamson, 1984). This comment indicates the potential difficulties of attempting to manage risk. Indeed, sometimes it seems as though it is not a science but more an art based on years of experience and 'gut-feelings'.

Risks and their interactions can emerge at anytime: at the front-end, during construction, or at operation stage and build-up to shatter carefully laid plans. Indeed, the only certainty is that unforeseeable events will materialize. Uncertainty springs up as issues are brought to the fore, dormant tensions emerge, and interdependent links are triggered (Miller and Lessard, 2000).

Sir Michael Latham considered that 'No construction project is risk free. Risk can be managed, minimized, shared, transferred or accepted. It cannot be ignored' (*Constructing the Team*, 1994). Everyone should be concerned with RM, because risk and uncertainty could have potentially damaging consequences on a project. For a property developer in the pre-feasibility stage it may lead to the question of whether to undertake a marginal project particularly if there is likelihood of the project finishing late and of over-budget. For a contractor or subcontractor, unforeseen risks may mean incurring losses that are not recoverable.

British Standard 4778 Section 3.1: 1991 defines RM as:

[T]he process whereby decisions are made to accept a known or assessed risk and/or the implementation of actions to reduce the consequences or probability of occurrence.

The process must have the aim of identifying and assessing the risks. RM and risk assessment as techniques will not remove all risks. The aim must be to ensure that risks are assessed and managed in an effective manner to achieve the overall objectives of the project.

The 1999, NAO report *Modernising Construction* highlighted inadequate use and understanding of value management (VM) and risk management (RM) as major barriers to improvement in construction performance as discussed in *Achieving Excellence in Construction Procurement Guide 04* (OGC, 2003).

There may be formal requirements for risk analysis (RA) for many reasons including: economic viability assessment, financial feasibility assessment, insurance purposes, accountability, contractual purposes, tendering, regulatory purposes and communication purposes (Cooper and Chapman, 1987).

Thomson and Perry (1992) identified that RM may involve:

- identifying preventative measures to avoid or reduce a risk;
- proceeding with a project stage-by-stage to reduce uncertainty though better information;
- considering risk transfer in contract strategy, with attention to the motivational effects and the control of risk allocation;
- considering risk transfer to insurers;
- setting and managing risk allowances in cost estimates, programmes and specifications;
- establishing contingency plans to deal with risks when they occur.

Traditionally risk in construction was either ignored or dealt with in an arbitrary way, for example, by including a 5% contingency factor in the estimate. Project contingencies provide an allowance to cover a client's risk exposure but make little contribution to its management; indeed this approach may contribute to the 'variation culture' (Rawlinson, 1999).

RM is an ongoing process throughout the life of the project, as risks will be constantly changing. RM plans should be in place to deal quickly and effectively with risks if they arise. It is important to work as an integrated project team from the earliest possible stages on an open-book basis to identify risks throughout the team's supply chains (OGC, 2003).

RM can be considered to have three stages: identification, analysis and response.

8.2   Risk identification

The initial step is the identification and assessment of the risks associated with a proposed construction project or contract at the early stages of the project's life. The identification process will form the basis whereby risks, uncertainties, constraints, policies and strategies for the control and allocation of risk are established.

Perry and Hayes (1985) suggest that the burden of responsibility for the identification of risks lies with the client, as they will be keen to achieve the overall objectives of completion within time, within budget and to an acceptable quality. The contractor will also need to be able to identify the risks in the contract in order to prepare the tender.

The potential risks in construction projects are many and varied. One of the most comprehensive lists of risks, identifying over 100 potential issues, was produced by Perry and Hayes (1985) classified into physical, construction, design, political, financial, legal-contractual and environmental.

Rawlinson (1999) identified that the principal categories of risk that the client may face resulting from a major capital project are potentially much wider than additional construction costs and could include the following (see also Table 8.1):

1. project risk – concerned mainly with time and cost;
2. consequential risk – the knock-on effects of project shortfalls on the client's business/ organization;
3. benefits risk – the effect of the project delivering more or less than the expected benefits;
4. The effect on share price or public perception of the business/organization due to public success or failures.

The above list is not of course definitive, indeed it would be foolhardy to think that it was. Some key issues are never considered as risk factors by client organizations. For example, on the Scottish Parliament building, the deaths during the early construction period of two key players – the First

Table 8.1 Sources of risk to client's business from construction projects.

Heading	Change and uncertainty due to
Political	Government policy, public opinion, change in ideology, dogma, legislation, disorder (war, terrorism, riots)
Environmental	Contaminated land or pollution liability, nuisance (e.g. noise), permissions, public opinion, internal/corporate policy, environmental law or regulations or practice or 'impact' requirements
Planning	Permission requirements, policy and practice, land use, socio-economic impacts, public opinion
Market	Demand (forecasts), competition, obsolescence, customer satisfaction, fashion
Economic	Treasury policy, taxation, cost inflation, interest rates, exchange rates
Financial	Bankruptcy, margins, insurance, risk share
Natural	Unforeseen ground conditions, weather, earthquake, fire or explosion, archaeological discovery
Project	Definition, procurement strategy, performance requirements, standards, leadership, organization (maturity, commitment, competence and experience), planning and quality control, programme, labour and resources, communications and culture
Technical	Design adequacy, operational efficiency, reliability
Human	Error, incompetence, ignorance, tiredness, communication ability, culture, work in the dark or at night
Criminal	Lack of security, vandalism, theft, fraud, corruption
Safety	Regulations (e.g. CDM, Health and Safety at Work), hazardous substances (COSHH), collisions, collapse, flooding, fire and explosion

Source: Godfrey, 1996.

Minister for Scotland, Donald Dewar and the Architect, Enric Miralles – had a significant impact on the project.

The CIRIA report 125 (Godfrey, 1996) states that 'It is impossible to identify all risks. To believe you have done so is counter-productive to RM and dangerous. Always expect the unexpected.'

It is worth reflecting on some of the unforeseen risks that have affected construction projects in the UK over the past 30 years:

- Power strikes/3-day working week/lack of materials, for example, steel (1970s);
- Widespread industrial action (1970s);
- 25% annual inflation (1970s);
- Civil unrest/miners strike (1980s);
- Poll-tax riots (1980s);
- IRA terrorism (1980s);
- Petrol-tax protestors (2000);
- Widespread flooding (2000 and 2007).

Likewise when considering risks likely to be encountered by contractors and specialists other risks may occur which are never anticipated, for example, a tower crane may collapse or the weather may make a significant impact. For example, on Foster + Partners' new high-rise Willis building in London, the wind stopped the tower cranes from operating for 40–50% of the time during the winter compared with the 20% anticipated (Lane, T., 2007).

Building magazine (16 February 2007) reported that Bovis Lend Lease had announced that it would change its tendering policy after taking a £48 million loss on work in Britain, primarily on its Manchester Joint Hospital PFI scheme. Bob Johnson, Bovis' global chief executive, said that 'the firm had taken it on without fully understanding the risks and did not price it correctly.'

Panel 8.1 Particular risks for main contractors and specialist contractors

- Poor tender/briefing documents;
- Client who will not commit;
- Inexperienced client;
- Non-standard contract documentation;
- Ultimate client failing to sufficiently acknowledge and reward quality and value for money;
- Poor design for construction, for example, when 'buildability' is not addressed;
- Unexpected problems relating to the site, such as contamination or unusual ground conditions;
- Co-ordination problems – this could be a particular problem for specialists;
- Component and/or materials suppliers unable to meet delivery and/or cost targets;
- Faulty components and/or materials;
- Accidents and injuries to staff;
- Weather interrupting work;
- Delayed payments;
- Poor documentation of records;
- Lack of co-ordination of documentation;
- Poor guidance for operatives;
- Poorly trained or inadequately trained workforce;
- Industrial disruption.

Source: *Constructing Excellence* 'Risk Management' Fact sheet (www.constructingexcellence.org.uk/ [accessed 20 July 2006]).

8.3   Risk analysis techniques

The purpose of risk analysis (RA) is to quantify the effects on the project of the risks identified. The first step is to decide which analytical technique to use. At the simplest level each risk may be treated independently of all others with no attempt made to quantify any probability of occurrence. Greater sophistication can be achieved by incorporating probabilities and interdependence of risks into the calculations but the techniques become more complex. The choice of technique will usually be constrained by the available experience, expertise and computer software.

Whichever technique is chosen, the next step requires that judgements be made of the impact of each risk and in some cases of the probability of occurrence of each risk and of various possible outcomes of the risk.

The main objective of RA must be to assess the effects on the project by the risks identified (see Table 8.2). The techniques range from subjective assessments through to the use of sophisticated techniques using computer software. The approaches can be categorized under two broad headings:

1. The summation of individual risk exposures to calculate a project risk allowance; techniques include expected monetary value (EMV) and Monte Carlo simulation;
2. The statistical calculation of average and maximum risk allowances; techniques include application of the central limit theorem and multiple estimating using the root mean square (RMS) method.

Identification of the potential risks can be achieved by:

- interviewing key members of the project team;
- organizing brainstorming meetings with interested parties;
- using the personal experience of the risk analyst;
- reviewing past project experiences.

Dr Steve Simister (2000) identified the differences between a qualitative and quantitative approach.

In the qualitative assessment, which is recorded and analysed in the 'risk register', it is necessary to ask the following questions: what is the risk? how might it occur? how likely is it? (probability) how good/bad might it be? (impacts) does it matter? what can we do? when should we act? who is responsible?

In contrast, in a quantitative assessment, the computer model, for example, a model developed based on spreadsheets and @Risk software is used for the following purposes: modelling uncertainty; simulating combined effects of risks; predicting outcomes, range min/max expected; testing scenarios; setting confidence limits; identifying criticalities and determining options.

Calculating risk allowances

Method 1: expected monetary value

The assessment identifies the impact of risks in terms of both the *impact* and the *probability of occurrence*. This can be expressed as the simple formula:

Risk exposure = Impact × Probability

It is important that all the potential risks and uncertainties which can affect the project and are likely to act as constraints on the project be identified as early as possible.

Once the risks have been identified, the risks are then subjected to an assessment that categorizes the risks into a subjective probability of occurrence and into three categories of impaction on the project – optimistic, most likely and pessimistic outcome. Two rules should be obeyed in the calculation: first, the most likely outcome must have the highest value, and second, the total value of probability for the three outcomes must always equal 1.

This method is simple and transparent and allows the consideration of more than one risk. However it has the disadvantage that it is unable to consider linkages between risks.

Consider the calculation of the risk allowance to be made for the potential increased lengths to the piling due to the uncertain ground conditions.

	Outcome £	Impact (I) £	Probability (P) %	(I × P) £
Cost plan allowance	1,350,000			
Optimistic outcome		150,000 (saving)	0.30	(45,000)
Most likely outcome		150,000 extra	0.50	75,000
Pessimistic outcome		250,000 extra	0.20	50,000
Expected monetary value				80,000

So £80,000 should be added to the cost plan allowance for the piling.

Depending on the size or complexity of the project, it may be necessary to carry out a *secondary* RA to identify consequential secondary risks.

This same technique can be used by contractors in order to establish an allowance for risk factors when compiling a tender or a quotation for a variation. David Neale, director of May Gurney (Construction) Ltd, described the difficulties in identifying the risks in a design and build (D&B)

highway project which involved widening and upgrading the existing carriageway. In his paper, Neale (1994) described the development of a crude risk model (similar to the above cost plan) in which the contractor calculated an explainable sum to be added to their tender.

In the event, none of the risks identified occurred, and instead the contractor encountered a totally unforeseeable event where the design responsibility was the contractor's and this cost more than the total allowance.

The author (Potts, 2003) also described a similar situation in connection with a quotation for a major variation submitted prior to the work being executed. The risk allowance proved inadequate to cover the substantial additional cost which was considered the contractor's risk once the quotation was accepted.

Decision trees

Decision trees (see Fig. 8.1) can be useful where the scenario is more complex. They are graphical representations useful in assessing situations in which the probabilities of particular events occurring depend on previous events, and can be used to calculate expected values in these more complex situations.

The decision tree shows two risks: A (Adverse weather at the contractor's risk) and B (Potential claim from the client of delay damages – acceleration is thus required to make up lost time).

Risk A has a 20% chance of occurring with a monetary value of £10,000. If outcome A occurs, a second risk B is introduced and there are three likely outcomes:

1. pay bonuses to own labour;
2. import additional labour;
3. subcontractor's responsibility.

The monetary value of Risk B is £30,000.
Using the decision tree the following financial risks are identified:

Outcome no. 1 has a financial risk of (£10,000 × 0.2) + (£30,000 × 0.25) = £9,500
Outcome no. 2 has a financial risk of (£10,000 × 0.2) + (£30,000 × 0.70) = £23,000
Outcome no. 3 has a financial risk of (£10,000 × 0.2) + (£30,000 × 0.05) = £3,500

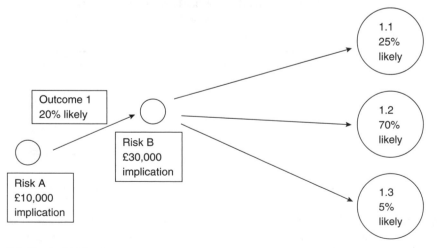

8.1 Simple decision tree.

So if possible you should try to achieve outcome no. 3 (subcontractor's responsibility!), as this has the least potential cost. This example shows how these calculations can easily become complex and highly theoretical. Some might argue that this conclusion could have been identified by inspection of the facts, common sense and gut feeling. Indeed this whole scenario may well rebound onto the main contractor if they try to implement harsh terms in the subcontract with the eventuality that the subcontractor's personnel walk off site.

Indeed this was the scenario in Hong Kong when the giant Gammon Kier Lilley JV (GKL) tried to impose tendered bills of quantities (BofQ) rates onto the local labour used by the architectural subcontractor. Due to a shortage of skilled workers the subcontractor was beginning to make a financial loss on the rates quoted in his tender and requested additional payments. GKL insisted that he should do the work at the rates quoted at tender and which were included in his subcontract agreement. The subcontractor's reaction was to take his men off site and the owner disappeared into China for two weeks. GKL soon realized who was carrying the risk – it was them, not their subcontractor!

Sensitivity analysis

The basis of a sensitivity analysis is to define a likely range of variation for elements of the project data. The final project cost or duration is then assessed for each variation in the data. In effect, a series of *what-if* estimates is produced.

The results of sensitivity analyses are often presented graphically, on a spider diagram, which readily indicates the most sensitive or critical areas for management to direct its attention towards.

One weakness of sensitivity analysis is that the risks are treated individually and independently. Caution must therefore be exercised when using the data directly to assess the effects of combination of risks.

Method 2: Monte Carlo simulation

Sophisticated analysis techniques are sometimes used to quantify the occasion of risks. Mathematical models and analytical techniques are often useful indicators of trends and problems. However these techniques should not be relied upon as the sole guide to the decision-making process.

The Monte Carlo simulation method relies on random calculation of values that fall within a specified probability distribution. The basic steps are as follows:

1. assess the ranges of variation for the uncertain data and determine the probability distribution most suited to each piece of data;
2. randomly select values for the data within the specified range taking into account the probability of occurrence;
3. run an analysis to determine values for the evaluation criteria for the combination of values selected;
4. repeat steps 2 and 3 above a number of times. The resulting collection of outcomes is arranged in sorted order to form a probability distribution of the evaluation criteria. The accuracy of the final distribution depends on the number of repetitions, or iterations, usually between 100 and 1,000.

Since the outcome from the Monte Carlo analysis is a collection of, say, 1,000 values of each evaluation criterion, it is unlikely that the same value for the evaluation criterion will be calculated more than a small number of times. The values are therefore grouped into class intervals. The results are presented as frequency and cumulative frequency distribution.

It is common to carry out a probabilistic time analysis with the aid of a CPM network to model the project schedule. The same method can be used for probabilistic cost analysis especially when the cost estimate can be broken down into the same categories or activities as the schedule and when cost risks are related to time costs (Norris, Perry and Simon, 1992).

Monte Carlo simulation enables linkages to be established between risks and is based on widely available computer software. However the calculations can easily become highly complicated.

Method 3: central limit theory

This is a simple technique used for calculating the overall risk allowance for projects, which provides a confidence limit of 90%. The model has three elements:

1. A base estimate which should be risk free;
2. A calculation of the individual risk allowances calculated using the formula: Risk allowance = Impact (I) × Probability (P) (the sum of these risk allowances should provide a 50% risk allowance);
3. Calculation of a 90% risk allowance using the formula:

$$X = 1.3 \sqrt{\Sigma(I^2 \times P \times (I - P))}$$

The 90% risk allowance is essentially a lump-sum contingency. The technique is easy to calculate on a spreadsheet and only one outcome needs to be considered; however, linkages between risks cannot be considered.

Method 4: multiple estimating using root mean square

This technique has been used extensively by the Ministry of Defence and other public sector bodies. The calculation requires the calculation of three estimates:

1. the base estimate which should be risk free;
2. the average risk estimate – defining the project contingency;
3. the maximum likely risk estimate – calculating the 90% risk allowance.

The risk estimate is derived from the formula: Risk allowance = Impact × Probability
The maximum risk allowance (MRA) is calculated using the following formula:

$$MRA = \sqrt{\Sigma \, ((I_{max} \times P_{max}) - (I_{ave} \times P_{ave}))^2}$$

(Where:
I_{max} = the maximum risk impact
P_{max} = the maximum probability, typically a 90% confidence limit
I_{ave} = the average risk impact
P_{ave} = the average probability, typically 50% confidence limit)
The calculation of both average and maximum risk allowances also requires the distinction between fixed and variable risks.

The multiple-estimating technique provides a thorough appraisal of the project risks. However the process requires a complex calculation requiring two estimates and the distinction between fixed and variable risks. The example above is based on an actual calculation used by the client's project managers/cost consultants when establishing the lump-sum contingency on a major complex building project in Birmingham.

Table 8.2 Project manager's risk analysis report at feasibility stage.

University of Metropolis
Project manager's report
Risk analysis – client project risks (feasibility stage)

Ref	Description of risk element	Type	Average risk			Maximum risk		
			Base value of risk element (£K)	Probability factor (F) or confidence limit (V)	Value (£K)	Probability factor (F) or confidence limit (V)	Value (£K)	Square of the deviation
	(a)	(b)	(c)	(d)	(e)	(f)	(g)	(h)
B010	Fitting out of school	F	70	0.9	63	1	70	49
B021	Teaching block 1	F	1,400	0.9	1,260	1	1,400	19,600
B023	Private finance initiative	V	N/A	50%	750	90%	1,500	562,500
B030	Equipment purchase	F	3,500	0.5	1,750	1	3,500	3,062,500
B043	Retail units main entrance	F	100	0.1	10	1	100	8,100
B046	Library	F	100	0.25	25	1	100	5,625
B050	Fixtures	V	N/A	50%	50	90%	100	2,500
B051	Existing boiler house	F	250	0.25	63	1	250	35,156
B061	Board room	F	T.B.A					0
				Average risk allowance	3,971	Square root of the sum of the deviations		1,923
						Add to the average risk allowance		3,971
						Maximum risk allowance (£K)		5,894

Notes
F = fixed (expressed as a ratio)
V = variable (expressed as a percentage)

If (b) is 'F' then (e) = (c) x (d) and (g) = (c) x (f)
If (b) is 'V' then both (e) and (g) are estimated costs reflecting the levels of confidence, that is, there is no 'Base Value' (N/A = not applicable).

Explanation of probability ratios for fixed risks:

Unlikely	0.10
Could happen	0.25
As likely not	0.50
Very likely	0.75
Almost certain	0.90
Certain	1.00

Explanation of confidence limits:
Average risk is always 50% confidence
Maximum risk is always 90%
confidence, that is, there is only 10% chance
of this being exceeded

8.4   Risk register

In the past decade or so the use of the risk register as a key control document (see Table 8.3) has gained acceptance with leading clients. The risk register lists all the identified risks and the results of their analysis and evaluation and information on the status of the risk. The risk register should be continuously updated and reviewed throughout the course of the project.

The UK Government's *Green Book, Appraisal and Evaluation in Central Government* identifies best practice for public sector bodies and notes that the risk register should contain the following information: risk number (unique within register); risk type; author (who raised it); date identified; date last updated; description; likelihood; interdependencies with other sources of risk; expected impact; bearer of risk; countermeasures; and risk status and risk action status (http://greenbook.treasury.gov.uk/annex04.htm [accessed 12 March 2007]). The risk register is important for the following five reasons:

1. monitoring and, if necessary, correcting progress on risk mitigation measures;
2. identifying new risks;
3. closing down expired risks;
4. amending risk assessment for existing risks;
5. approving the drawdown of project contingencies by the client when required.

The author, Potts (2003), identified that the key to the successful management of risk throughout the project cycle on the Severn Trent Derby Sewage Works project was the compilation of a joint contractor/client risk register with regular reviews and modifications. This D&B target-cost project with a pain/gain arrangement was carried out under a five-year framework agreement. Under this project, the first attempt at identifying the risks was based on an unstructured methodology without constraints. The resulting risks were then listed on the risk register; the process was repeated with the aid of a risk matrix, checklists and past experience.

The *Contractor/Client joint risk register* was a dynamic control document which was evolved throughout the life by Seven Trent Water (STW) and the contractor during the life of the project. All the risks, including potential variations, were subject to continuous review and modification.

8.5   Risk response

The response to the risk will be appraised by the severity of the risk to the project. There are four risk mitigation strategies that can be adopted by the client and project team in order to reduce the risk exposure associated with a project: avoidance, reduction, transfer or retention.

Avoidance

If the situation is assessed whereby the risk is judged to have such a serious consequence, then the situation may warrant a reappraisal of the project. It may be necessary to review the project's aims, either to reappraise the concept or to cancel the project.

Reduction

Reducing the risks may involve redesigning the project, changing the procurement strategy or undertaking additional soil investigation to minimize changes to the foundations, changing the specification or incorporating different methods of construction to avoid unproven construction techniques.

Table 8.3 Typical contractor/client joint risk register (part only).

	Description of risk	Description of impact	Impact (£K)	Probability (%)	Owner	Included in target	Sum included in target	Comments
A	General ground problems	Disposal costs of unforeseen contaminated material	20	50	50/50	Yes	£10K	200 m³ max assumed
B	Weather	Adverse weather – additional cost and time	50	50	50/50	Yes	£25K	Additional temporary covers
C	External restraints	Land purchase on section – legal time delay	100	5	Client	No	£5K	Non-critical section
D	Contractual problems	Delay of framework suppliers	40	10	50/50	Yes	£4K	Consider acceleration
E	Design/survey	Removal of asbestos in existing structures	9	30	Client	No	£2.7K	Specialists required
F	Operational requirements	Delay in issue of permits	20	10	Client	No	£2K	1 week assumed
G	Site specifics	Site security – loss of equipment	50	50	50/50	Yes	£25K	Check details of insurance
H	Price build-up	Increased cost – Inflation	50	50	50/50	Yes	£25K	Steel/cement prices volatile

Source: Potts, 2003.

Panel 8.2 Risk registers

Details of the risks should appear in the risk register. However the detailed nature of the risk register means that it can be difficult to capture a meaningful summary of the current exposure to risk. A suggested approach is to assess each risk against a matrix of probability (high/medium/low probability) and impact (high/medium/low impact).

Risks that have a high score on both probability and impact, or a high score on one and medium on the other, are assigned a Red status (high risk requiring careful attention); those that are low probability and low impact, or low on one and medium on the other, are given a Green status (risks need watching but not a priority concern). Regular recording of the progress of risks is essential, because their status can change rapidly. A summary of Red risks could be reported at senior level via project board meetings, for example, so as to concentrate on the areas of highest risk.

Source: *Achieving Excellence in Construction* (2003), Procurement Guide '04.

Transfer

Perry and Hayes (1985) identified four common routes for the transfer of risk:

1. client to contractor;
2. contractor to subcontractor;
3. client, contractor, subcontractor or designer to insurer;
4. contractor or subcontractor to surety.

However it should be noted that implementing transfer of the risks to others may result in development of different contract terms, payment of higher fees or additional premiums.

If risks can be transferred, their consequences can be shared or totally carried out by someone other than the client. The client will be expected to pay a premium for this, so responsibility for initiating this form of risk response lies with the client. Abrahamson (1984) considered that a party should bear a construction risk where:

1. it is in his control, or
2. the party can transfer the risk by insurance, or
3. the preponderant economic benefit of running the risk accrues to him, or
4. to place the risk on him is in the interests of efficiency, or
5. if the risk eventuates, it is not practicable to transfer the loss to another.

Retention

Risks that are retained by either party may be controllable or uncontrollable by that party. Where control is possible it may be exerted to reduce the likelihood of occurrence of a risk event and also to minimize the impact if the event occurs. It will be necessary to include a project contingency fund.

8.6   Strategic risk management

A most significant review of large engineering projects – defined as airports, urban transport, oil fields and power systems – was undertaken by an international team lead by Roger Miller, the Hydro-Quebec/CAE Professor of Technology Management at the University of Quebec at Montreal. The purpose was to benchmark 60 worldwide projects and identify best practice, to shape research issues and to share learning (Miller and Lessard, 2000).

The key findings from observing how sponsors wrestled with the risks inherent in projects, as well as those arising from possible conflicts among the various co-specialized partners or stakeholders, are that the lead sponsors have developed a strong repertoire of strategies for coping with risk. The ability to frame risks and strategies represents a core competence for them. This competence spans the five types of management philosophies:

1. obtaining and framing information;
2. designing a process with a long *front-end* before technical, financial and institutional details are locked in, followed by rapid execution of the project;
3. building coalitions that bring together varying information and skills and are structured to create strong incentives for performance and mitigate conflicts of interest;
4. the allocating of risk to the party best able to bear it;
5. transforming institutional environment risks through the creation of long-term coalitions that incorporate powerful influences on laws and rules.

These recommendations are very much in line with the OGC advice and the *Constructing Excellence* programme within the UK.

Fig. 8.2 has been developed based on a letter sent to *Building* magazine by Tony Clarke, Director of Management Services at Knowles, in July 2005. In this letter he listed the causes of disputes on projects such as the Millennium Stadium, the Scottish Parliament building, The Great

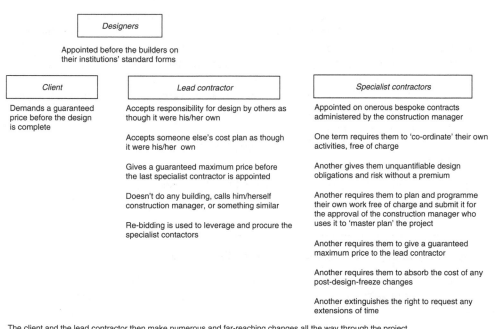

8.2 The real causes of disputes, or who carries the risk? (source: developed based on letter from Tony Clarke to *Building* magazine, 22 July 2005).

Eastern Hotel, The British Library, The Jubilee Line, The Royal Brompton Hospital and Portcullis House, and conjectured that these same issues may be the cause of the problems on the Wembley Stadium project.

This is a most important contribution to our understanding of RM on major projects. It shows that in many instances clients have attempted to dump the risk onto the lead contractor who in turn has passed the risk down to the specialist contractors. The case studies throughout this book demonstrate that in practice this approach does not work and may rebound onto the client with catastrophic consequences.

The words of Sir John Egan 'We should all rethink construction' ring somewhat hollow. It is necessary to go back to Sir Michael Latham's recommendations on the 13 essentials of a modern contract contained in *Constructing the Team* and embrace the philosophy of partnering, mutual trust and co-operation found in the NEC Engineering and Construction Contract.

8.7   Case studies

Case study 1: the Millennium 'wobbly' Bridge, London

The Millennium Bridge over the Thames, linking the newly opened Tate Modern Gallery at Bankside with the City of London at St Paul's, was the first pedestrian bridge built over the Thames for over one hundred years.

The bridge was intended to be one of the landmark projects heralding the new millennium. The innovative and complex structure that featured a 4 m wide aluminium deck flanked by stainless steel balustrades and supported by cables was designed by a joint venture comprising architect Norman Foster, sculptor Anthony Caro and structural engineers Ove Arup.

Such was the interest in the new bridge that when it opened to the public on 10 June 2000, an estimated 80,000 to 100,000 people crossed it. It soon became clear that all was not well as the deck swayed about and many reported feeling seasick.

After a prolonged series of tests, it was decided to adopt passive damping system which would harness the movements of the structure to absorb energy.

After nearly two years of testing, the alterations were deemed a success and the bridge finally opened to the public in February 2002. The alterations had cost an extra £5 million on top of the initial £18 million.

Postscript: The risk of design failure was not one of the ten principal risks identified in the pre-contract RM exercise!

Source: www.urban75.org/london/millennium.html (Rawlinson, S., 1999).

Case study 2: Royal Opera House, Convent Garden, London

The £200 million refurbishment of the Royal Opera House was one of the largest and most complex projects funded by the National Lottery Board. The project was executed using the construction management approach as this was considered an ideal process for managing risk.

The aim was to identify risks at the earliest possible opportunity and proactively manage them through the risk register. Actions were taken to contain and reduce the risks and to transfer the risks remaining through the contract to consultants and trade contractors. Project risks were therefore only transferred to trade contractors after they had been identified, analysed, quantified and minimized. The risks were clearly described in the tender documents and specific actions agreed with the trade contractors.

Cost risks on the project were minimized by the early award of those key trade packages, which have a large element of specialized detailed design, on a lump-sum design and construct basis. Around 70% of the total project value was procured prior to project construction; the remaining 30% was the subject of a detailed cost estimate.

Rigorous cost-control procedures were established to identify and obtain client approval to all post-contract variations before implementation.

Time risks on the project were minimized by detailed analysis of the programme, construction methods and sequences, resources and logistics. Methods included critical path analysis, resource analysis and resource levelling, 3D CAD modelling of construction phasing and construction programme. Detailed method statements were produced for each critical element and comprehensive RA undertaken of the programme linked to probability simulations

Source: Trotter, 1995.

Case study 3: RM on the Scottish Parliament Building

Risk management for the Holyrood project was not good practice. In my 2000 report I concluded that accounting for risk was insufficient. I showed that contrary to good practice there was no quantified allowance for the major risks facing the project. I recommended that this should be established and the results used as a basis for an action plan to manage the risks. Project management introduced a process for quantifying risks from October 2000 and then conducted a number of risk reviews.

Although by definition risk is uncertain, some 70% of the risk identified by the October 2000 workshop was for items that were categorized as 'highly likely' i.e. assessed by the workshop members as having at least a 95% probability of occurring. Each of these items carried with them varying levels of likely impact on programme.

Moreover the risk evaluation did not attempt to evaluate the monetary value attached to the risks to the time schedule. If it had it would have added further risk cost.

However, in the Holyrood project the general approach was to accept cost increases and include them in the forecasts as the risks materialized. Because there was no agreed budget limit after June 2001, there is little evidence that forceful action was taken to prevent or reduce the increase in cost.

Source: Auditor General for Scotland, Audit Scotland (2004).

8.8 Conclusion

We have reviewed some of the main techniques and methodologies for RM relevant for construction projects and identified that real benefits can occur with their use. However there is no panacea for successful management of risk; it should be viewed constructively and creatively. Rigid application of a set technique or procedure is not advocated or encouraged. Indeed methodologies are, relatively speaking, in their infancy and evolving with practice.

Interest in RM comes mainly from educated clients and is regularly adopted as integrated front-end service. Ongoing RM studies throughout the project are largely limited to the public and utilities sectors. Wood and Ellis's research found that the use of RM workshops and the production of risk registers is commonplace. The use of Monte Carlo simulation is widespread through the use of specialist software. However there seems some scepticism among the leading UK cost consultants regarding the usefulness of complex RA techniques and there is a predisposition to rely on judgement based on experience (Wood and Ellis, 2003).

The initiative for the application of RM lies with the client and their professional advisers, particularly the project manager and cost manager. Some changes in normal policy may be required, for example, building adequate time for RM into the project programme, training and perhaps experimentation with techniques.

8.9   Questions

1.   Identify two unexpected events that have occurred on a construction project with which you have been involved. What was the impact on the project and how were they dealt with?
2.   The University of Metropolis, together with the local City Council, is developing a brownfield site adjacent to an old gasworks in order to provide a large Science and Technology Park. It is anticipated that the project will be carried out in phases over a period of five years.

    a.   Identify the main strategic risks for the client organization and how these might be avoided/mitigated.
    b.   Identify the main risks for the contractor and how these might be avoided/mitigated.

3.   John W. Hinchey, partner with King and Spalding LLP in Atlanta, Georgia, identified *Ten ways owners can avoid or mitigate construction risk*. Critically review this article making specific reference to your own market. (www.rics.org/Management [accessed 10 March 2007])

Bibliography

Abrahamson, M. (1984) 'Risk management', *International Construction Law Review*, vol. 1, no. 3, pp. 241–264
Audit Scotland (2004) 'Management of the Holyrood Building Project', prepared for the Auditor General for Scotland, June, Audit Scotland
Clarke, T. (2005) 'Fifteen golden rules', letter to *Building* magazine, 22 July, p. 32
Cooper, D. and Chapman, C. (1987) *Risk Analysis for Large Projects: Models, Methods and Cases*, John Wiley & Sons
Edwards, L. (1995) *Practical Risk Management in the Construction Industry*, Thomas Telford
Flanagan, R. and Norman, G. (1993) *Risk Management and Construction*, Blackwell Scientific
Godfrey, P.S. (1996) *Control of Risk: A Guide to the Systematic Management of Risk from Construction*, Special Publication 125, Construction Industry Research and Information Association (CIRIA)
Hill, E. (1998) *Managing Risk – Methods for Software Systems Development*, Addison Wesley Longman
Institution of Civil Engineers & the Faculty and Institute of Actuaries (1998) *Risk and Management for Projects*, Thomas Telford
Lane, T. (2007) 'Dancing with disaster', *Building* magazine, 27 April
Latham M. (1994) *Constructing the Team*. Final Report of the Government Industry Review of Procurement and Contractual Arrangements in the UK Construction Industry, HMSO, London
Miller, R. and Lessard, D.R. (2000) *The Strategic Management of Large Engineering Projects*, The MIT Press
Neale, D. (1994) 'New Highways Design & Build Contract: What the Contractor Thinks about the Contract', I.H.T. Cambridge 1994 symposium

Norris C., Perry, J.G. and Simon, P. (1992) *Project Risk Analysis and Management. A Guide by the Association of Project Managers*, CPS Project Management Ltd

OGC (2003) *Achieving Excellence in Construction Procurement Guide 04: Risk and Value Management*, OGC

Perry, J.G. and Hayes, R.W. (1985) 'Construction projects – know the risk', *Chartered Mechanical Engineer*, February, pp. 42–45

Potts, K. (2003) 'Risk Management on Variations – Two Civil Engineering Case Studies', RICS Foundation Construction and Building Research Conference, University of Wolverhampton, 1–3 September

Raftery, J. (1994) *Risk Analysis in Project Management*, E&FN Spon

Rawlinson, S. (1999) 'Risk Analysis and Risk Management', Notes to support Television Education Network Video, January

Simister, S. (2000) 'Risk Management', lecture to postgraduate students, University of Wolverhampton, unpublished notes

Smith, N.J. (1999) *Managing Risk in Construction Projects*, Blackwell Science

Thompson, P. and Perry, J. (ed.) (1992) *Engineering Construction Risks: A Guide to Project Risk Analysis and Risk Management*, Thomas Telford

Trotter, S. (1995) 'Procurement – a Case Study', Effective Project Management Conference, School of Business & Industrial Management, 4 December

Wood, G.D. and Ellis, R.C.T. (2003) 'Risk management practice of leading UK cost consultants', *Engineering, Construction and Architectural Management*, vol. 10, no. 4, pp. 254–262

Useful websites

www.constructingexcellence.org.uk
www.greenbooktreasury.gov.uk/annex04.htm

9 Whole-life costing

9.1 Introduction

Whole-life costing (WLC) is now established as an important tool which is changing the whole approach to design, procurement and construction and delivering major benefits. WLC is used to describe a form of modelling technique which can embrace a mixture of capital and running costs. *Life-cycle costing* is an older term for the same thing; *costs in use* is now an obsolete term.

The New Construction Research and Innovation Strategy Panel (nCRISP) defines WLC as 'the systematic consideration of all relevant costs and revenues associated with the acquisition and ownership of an asset' (Constructing Excellence, 2004).

The Building Research Establishment gives the definition: 'Assessment of the whole life performance and cost of an asset over its lifetime takes into consideration initial capital costs and future costs, including operational costs, maintenance costs and replacement/disposal costs at the end of its life.'

Many public and private sector clients now procure on cost of ownership, not on capital cost. Recent initiatives such as *Achieving Excellence* and the drive for *Egan compliance* among housing associations demonstrate this trend. Local authorities often adopt WLC as part of their response to their duty to deliver best value.

Consortia formed to undertake private-finance initiative (PFI) and Private-Public Partnership (PPP) also demand identification of whole-life costs in order to prepare detailed financial and risk management plans (RM) for projects. PFI and PPP contracts include non-availability clauses that may impose severe financial penalties on consortia running schools if, say, a maintenance problem leads to a classroom being unavailable for use. So choices about a roof finish should not just be considered in terms of installation and maintenance costs – the whole-life costs should take into account the risk of the roof leak, the cost of repairs and associated penalty.

In the report 'Better Public Building' (produced in 2000 by the Department of Culture, Media and Sport) the then prime minister Tony Blair states 'integrating design and construction delivers better value for money as well as better buildings, particularly when attention is paid to the full costs of a building over its whole lifetime.'

WLC is not an optional extra. The Treasury, the National Audit Office and the Audit Commission expect it. A new British Standard BS ISO 15686 – *Service Life Planning of Buildings and Constructed Assets* – provides the foundation for service-life planning and the use of whole-life costing.

9.2   Understanding the relevance of whole-life costing (WLC)

Whole-life costs are substantially greater than capital or initial costs – it is estimated that the operational expenditure will be 5–10 times as much as the capital cost. However these ratios are small when compared with the ratio of capital expenditure to the operating costs of businesses occupying the building which could be anything between 100 and 200 times as much as the their building's initial costs. These ratios indicate that a 1% improvement in productivity/output of staff would effectively pay the entire capital costs of the building (*Constructing Excellence*, 'Whole Life Costing Fact Sheet', www.constructingexcellence.org.uk/pdf/fact_sheet/wholelife.pdf [accessed 10 April 2007]).

The following are considered to be some of the major benefits in implementing WLC:

- Encourages communication between the stakeholders and leads to an improved project definition;
- Clarifies the cost of ownership and occupation;
- Optimizes the total cost of ownership/occupation;
- Enables early assessment of risks;
- Promotes realistic budgeting;
- Encourages discussion and decisions about materials choices;
- Enables best value to be attained;
- Provides actual figures for future benchmarking.

WLC should be included in the client's brief. It should be used as a decision-making tool throughout the procurement, construction and use of the project stages; for example, during initial investment appraisal, feasibility study of alternatives, outline and detailed design, tender appraisal, assessment of variations, handover and post-occupancy.

WLC has the potential for adding real value to a project. However it is critical to involve the whole supply chain early in the design cycle, as 80% of the future costs of running maintenance repair is fixed in the first 20% of the design process. Experts in building services and facilities management should not be overlooked during the early design assessment if the full long-term environmental and economic advantages are to be secured.

The concept of WLC was first introduced into the procurement assessment on the Ministry of Defence's *Building Down Barriers* projects. As part of the Defence Estates Prime Contracting Strategy, potential contractors who bid had to forecast the whole-life costs along a series of milestones. Part of the payment on the seven-year project was based on performance against the milestones.

9.3   The basic steps in whole-life costing (WLC)

WLC is one of three evaluation processes that need to be undertaken during the design phase of any project. The other two are technical evaluation and environmental evaluation. The final choice of scheme will be a compromise between these three. Before a WLC analysis can be undertaken, it is essential that a value engineering (VE) exercise is undertaken, and repeated as necessary, in order to remove all unnecessary costs from the functional/technical specifications.

The Constructing Excellence fact sheet *Whole Life Costing* identifies that the following four basic steps should be followed in order to identify the whole-life cost of an asset:

1.   identify the capital and operational costs and incomes;
2.   identify when they are likely to occur;

3. use discounted cash-flow analysis to bring the costs back to a common basis – items should normally be entered into the analysis at the current cost and a *real* (excluding inflation) discount rate applied. Normally this will be done on a commercial spreadsheet package, which includes equations for discounted cash flow;
4. undertake sensitivity analysis of the variables such as the discount rate, the study period, the predicted design lives of components, assumptions about running costs etc.

Expected life of a component

Predicting the life of a component is not an exact science. Numerous factors interact to determine the durability in practice. For example, there may be several levels of different specification within one component. Furthermore the actual replacement interval is often determined by economic, technical or functional obsolescence. It is notoriously difficult to assess when or how obsolescence might strike. Critical factors including changing land values on which the building stands, changing IT infrastructure/cabling requirements and changing safety requirements might contribute in rendering certain buildings obsolete.

Feedback from practice is an important source of durability data and includes the following:

*   latent defects insurance schemes (e.g. NHBC, Zurich);
*   condition surveys and defects investigations;
*   maintenance records/repairs databases.

Published information on the life expectancy of building components is also available from the following sources:

*   RICS Building Maintenance Information (BMI) a partner to the Building Cost Information Service (BCIS) and the Building Research Establishment (BRE);

Panel 9.1 Determining element life

Kirk and Dell'Isola (1995) identified three different definitions of the life of equipment, materials or other components.

Economic life – estimated number of years until that item no longer represents the least expensive method of performing functions required of it.
Technological life – estimated number of years until technology causes an item to become obsolete.
Useful life – estimated number of years during which an item will perform the functions required on it in accordance with some pre-established standard.
These headline statements are meaningless without some form of context. For example, the same water pump is unlikely to have as long a life when used in a hard water area compared to one maintained exactly the same way in a soft water area.

Source: *Elements needed for a Whole Life Cost Analysis* (www.bsria.co.uk/press/?press=345 [accessed 2 June 2007]).

- Housing Association Property Mutual (HAPM) *Component Life Manual* (E&F Spon, 1992);
- Building Performance Group (BPG) *Building Fabric Component Life Manual* (E&F Spon, 2000);
- Building Performance Group (BPG) *Building Services Component Life Manual* (Blackwell Scientific, 2000);
- Chartered Institution of Building Services Engineers (CIBSE) *Guide to Ownership, Operation and Maintenance of Building Services* (CIBSE, 2000).

Another useful resource for component durability is The BLP (Building Life Plans Ltd) *Construction Durability Database* (www.componentlife.com [accessed 3 September 2007]). This database is funded by the Housing Corporation which provides up-to-date durability and specification data for 900 building components and is based on a 15-year data collection and research. Access to the database and to the updated information is provided free of charge to employees of Registered Social Landlords.

Elements to consider

The following items should be considered in the whole-life cost model:

1. Lifespan of building or asset;
2. Construction: site, design, construction, health and safety, commission, fit-out, professional fees, in-house fees, statutory fees, finance etc.;
3. Facility: rent/rates, energy for heating/cooling/power/lighting, utilities, maintenance, repair/replace, refurbish, management, security, cleaning etc.;
4. Disposal: dismantle, demolition, sale etc.

The list is not comprehensive and will depend on the type of building or asset. However it gives some indication of the key issues to consider.

9.4   Money, time and investment

A sum of money received some time in the future will always be of worth that is less than the same sum of money today and the difference will depend on the following:

- the length of time involved;
- future risks;
- the probable interest rates.

Panel 9.2 Establishing maintenance and energy costs

Over the life of the analysis, the combined revenue cost of utilities and maintenance can easily exceed the original capital expenditure. For example, it has been estimated that a highly serviced healthcare facility could spend the maintenance and operations budget equivalent of the capital cost every 3–5 years.

Relying on rule of thumb data can lead to a high degree of inaccuracy. The maintenance costs on 'Hospitals' vary between a lower cost of £0.91/m² and an upper cost of £36.46/m² (1:40); while the difference on 'Offices' varies between £0.09 and £35.89/m² (1:400)!

Source: *Elements needed for a Whole Life Cost Analysis* (www.bsria.co.uk/press/?press=345 [accessed 2 June 2007]).

In doing the calculations it is a good idea to assume an interest rate that would reflect likely inflation and any special risks over the period concerned rather than a rate, which might actually be obtainable today.

In considering development finance we have three kinds of expenditure/income which we need to compare with each other:

- lump sums today;
- lump sums in the future;
- sums of money occurring at regular intervals during the period under consideration.

We cannot compare these, one with the other, unless we modify them in some way to put them on a common basis. There are two basic methods, and as usual they are just different ways of expressing the same thing – *present-day value* and *annual equivalent*.

Present-day value

All expenditure is expressed as the capital sum required to meet present commitments plus the amount which would have to be set aside today to provide for future payments, discounted to allow for accumulation of interest. Income is similarly treated; future income is discounted to the present day in the same way.

Table 9.1 shows that £1,000 received in one-year time (based on 5% interest rate) will have the present value of £952. Likewise, if £1,000 is received in a 10-year period, it will have a present value of only £614.

The concept of Net Present Value (NPV) – value at today's date – is important as it is used by major international clients, including the World Bank and the Hong Kong Mass Transit Railway Corporation, to evaluate contractors' bids. Thus if a contractor's lump sum tender was low but they require a greater proportion of payment in year 1, this would be reflected in the NPV calculation.

The UK government's recommended discount rate is 3.5%. Calculating the present value of differences between the stream of costs and benefits provides the NPV of an option. The NPV is the primary criterion for deciding whether government action can be justified (HM Treasury, 2003).

Table 9.1 Present value and discount rate (based on 5%).

Time (mid- year)	0 (£)	1 (£)	2 (£)	3 (£)	4 (£)	5 (£)	6 (£)	7 (£)	8 (£)	9 (£)	10 (£)
PV of payment (mid-year)	1,000	952	907	864	823	784	746	711	677	645	614

Panel 9.3 Observations on the discount rate

The change from 6% to 3.5% for the standard rate to be used (dictated by HM Treasury in April 2004) effectively puts a higher weight on future costs, with the aim of encouraging longer-term more sustainable development.

The choice of the discount rate (interest rate) used can have a dramatic effect on the outcome of the analysis. As an example, an annual energy bill of £100,000 over 30 years will have a present value of around £1.84 million if a 3.5% interest is taken, but only £656,600 at 15%.

Annual equivalent

Annual equivalent is the total of the following:

- any regular annual payments and income, such as wages, rents etc.;
- annual interest on items of capital expenditure;
- a sinking fund – the amount which would have to be put away annually to repay the capital cost at the end of the period.

Alternatively, the annual interest and sinking fund can be combined and expressed as the annual instalments which would be required to pay off the capital costs and interest over the term of years in question (rather like paying off a loan for a house through a mortgage).

Both of these methods will give a similar answer, and which one is used is purely a matter of convenience and depends on whether you are thinking in terms of capital finance or of annual income and expenditure.

9.5   Calculations

In the following formulae n represents the number of periods and i is the interest rate expressed as a decimal fraction of the principal, for example 5% = 0.05.

The following standard time value of money tables should be downloaded from the Internet (www.studyfinance.com/common/table3(table 4).pdf [accessed 10 January 2008]).

- Table 1 – Present Value Factors
- Table 2 – Present Value of Annuity Factors.

  Formula 1 – Compound Interest $= (1 + i)^n$

If a sum of money is invested for some years it will have earned some interest by the end of the first year. Compound interest assumes that this earned money is immediately added to the principal and reinvested on the same terms, this process being repeated annually.

What will be the value of £5,500 invested at 9% compound interest for 5 years?

Formula $(1+i)^5$ and shows that £1 so invested will grow to £1.54.
£5,500 will grow to £5,500 $\times$ 1.54 = £8,470.
Formula 2 – Present value of £1 $= 1 / (1 + i)^n$

In the compound-interest example, £5,500 invested for 5 years at 9% grew to £8,470. Conversely, the present value of £8,470 in 5 years' time at 9% interest is £5,500, that is, the amount that will grow to that sum at the end of 5 years.

What is the present value of £1,200 in 35 years time discounted at 10% per annum?

By the use of the formula the present value of £1 in such circumstances is 0.0356 (refer table 1 – Present Value Factors). The present value of £1,200 is therefore 1,200 $\times$ 0.0356 = £42.72.

Formula 3 – Present value of £1 payable at regular intervals
$[(1 + i)^n - 1] / [i(1 + i)^n]$

This formula shows the present value of future regular periodic payments or receipts over a limited term of years. It is, therefore, very useful for assessing the capital equivalent of things such as running costs, wages or rents.

What is the present value of £1,200 payable annually for 10 years assuming an interest rate of 8% per annum?

Example: it is desired to compare the whole-life costs of two different types of windows (type A and type B) to an office building, whose life is intended to be 40 years. The rate of interest allowed is 3% per annum compound.

Whole-life costs of windows – type A

Windows type A will cost £900,000; will require redecorating every 5 years at a cost of £20,000 and will require renewing after 20 years at a cost of £1,200,000.

	£
Initial cost	900,000
Present value at 3% of the following:	
Redecoration after 5 years £20,000 @ 0.8626 (present value of £1)	17,252
Redecoration after 10 years £20,000 @ 0.7441	14,882
Redecoration after 15 years £20,000 @ 0.6419	12,838
Renewal after 20 years £1,200,000 @ 0.5537	664,440
Redecoration after 25 years £20,000 @ 0.4776	9,552
Redecoration after 30 years £20,000 @ 0.4120	8,240
Redecoration after 35 years £20,000 @ 0.3554	7,108
	£1,634,312

Whole-life costs of windows – type B

Windows type B will cost £1,250,000 and will last the life of the building without any maintenance, although a sum of £300,000 is to be allowed for general repairs after 20 years.

	£
Initial cost	1,250,000
Present value at 3% of the following:	
Repairs after 20 years £300,000 @ 0.5537	166,110
	£1,416,110

Saving by using Windows type B is therefore £1,634,312 – £1,416,110 = £218,202. It would therefore appear to be justifiable to use the initially more expensive type B windows, as this will prove much cheaper in the long run.

Inflation

The discount rate is not the inflation rate but is the investment *premium* over and above inflation. Provided inflation for all costs is approximately equal, it is normal practice to exclude inflation effects when undertaking life-cycle cost analysis (OGC, 2003). In recent years, the annual inflation in the UK has been 2–4% per annum; however ten years ago it was in double figures and in the early 1970s following the Middle East crisis was over 25% per annum.

The solution to the example above did not take into account inflation. If the original windows were worked out on the basis of 10% annual inflation and 13% interest we would get much the same result as with no inflation and interest at 3%.

9.6   Problems with assessing whole-life costs

We have seen how WLC enables us to consider the long-term implications of a decision and provides a way of showing the cost consequences. The eminent researchers and authors Douglas Ferry and Peter Brandon identified that there are a number of potential fundamental problems in using a WLC approach (Ferry & Brandon, 1999), such as:

1. Initial and running costs cannot really be equated:

    a.   The maintenance charges will fall upon the purchaser not on the developer;
    b.   Even with public buildings, for example, schools, the bulk of the construction costs are paid for by one authority with another authority responsible for maintenance;
    c.   Money for capital developments is often more difficult to find than money for current expenditure;
    d.   Hardwearing materials may give an old-fashioned appearance and may be replaced before they are life expired.

2. The future cannot really be forecast:

    a.   The cost of maintenance is pure guesswork;
    b.   The amount of money spent on decoration and upkeep is determined more by the body responsible for maintenance; for example, new owners than by any quality inherent in the materials;
    c.   Major expenditure on repairs is usually caused by unforeseen failure of detailing, faulty material or poor workmanship and is almost impossible to forecast;
    d.   Interest rates cannot be forecast with any certainty, particularly over long periods. Would you like to guess what the Bank of England (or the European Bank) would do in say 20 years?

These comments seem to have been written prior to the introduction of PFI projects, which would seem to negate several of the above observations.

More recently, the National Audit Office Report *Improving Public Services through Better Construction* (NAO, 2005) identified four key barriers to successful WLC. First, the lack of clarity on what is meant by WLC. Second, a lack of robust historical data on running costs. Third, people making investment decisions need a tool not just based on cost but on other drivers such as time, sustainability, quality and return on investment. The calculations are done in a vacuum and there is no way of comparing and evaluating the options. Finally, there is a lack of tangible evidence of the benefits of WLC (Green, 2005).

In an effort to address some of these issues the Building Cost Information Service (BCIS) has developed a standardized approach to WLC. In 2007, BCIS launched *BCIS Occupancy Online*. Initially this service would provide information at the building level based on the estimates of maintenance and occupancy based on a range of building types and a profile for that expenditure based on the BMI occupancy cost plans. The user would be able to change the time period and the inflation or discount rate and adjust the costs for time and location to provide cash flow, at current or future prices and at NPV (Martin, 2007).

9.7   Whole-life value (WLV)

Increasingly major procurement in the public and private sector is being undertaken on the basis of not just lowest capital, or even whole-life costs, but *value* (Bourke *et al.*, 2005). Whole-life value (WLV) encompasses economic, social and environmental aspects associated with the design, construction, operation, decommissioning and where appropriate the reuse of the asset of its constituent materials at the end of its useful life.

An important part of WLC is compiling the life-cycle assessment (LCA). This is a systematic set of procedures for compiling and examining the inputs and outputs of materials and energy and the associated environmental impacts directly attributable to the functioning of a product or service system throughout its life cycle.

However WLV includes more than WLC or LCAs, which are integral to the process. The application of WLV includes the consideration of the perceived costs and benefits of some or all of the stakeholders' relevant value drivers. The key techniques that are integral to WLV evaluations of building and infrastructure projects include:

- *WLC and LCA*: WLC deals primarily with financial costs, whereas LCA deals primarily with environmental impacts.
- *Multi-criteria analysis (MCA)*: MCA is used in conjunction with both WLC and LCA to evaluate alternative options based on criteria developed with stakeholders.
- *Group decision-making processes*: these processes include value management and risk management to engage stakeholders in the WLV process.

The WLV process will involve a number of iterations between the various stages and can be tailored to fit in with Gateway reviews such as the OGC *Gateway Process*.

The publication *Achieving Whole Life Value in Infrastructure and Buildings* (Bourke *et al.*, 2005) is a significant landmark in the subject of WLV. It has been compiled by eminent academics and practitioners and identifies how WLV is achieved, the techniques and methods thereof and includes four case studies. It should be recommended reading for all construction cost consultants.

9.8   Conclusion

The introduction of PFI contract requirements and high running costs has meant that WLV is gaining more acceptance within the construction/property sector. WLV is used early in the project cycle for feasibility and investment cases and for calculating budgets for maintenance.

The objectives of WLV are admirable, but in the past it often failed to deliver what it promised due to the absence of appropriate data. However more data is now becoming available and being incorporated into sophisticated models. For an excellent example of a whole-life cost model for two call centres, see Dudley and Derby, developed by Citex (now called Bucknall-Austin) see *Building* magazine, 23 July 1999, pp. 72–79.

With the development of the concept of WLV, which embraces not only costs but also environmental issues, value management and risk management, Kathryn Bourke at the BRE and colleagues have created a powerful new technique for the future.

9.9   Questions

1.   Consider the typical cash flow over a 50-year life expectancy of the following components:

   a.   Flat roofs: two-coat built up felt/asphalt;
   b.   Floor finishes: carpet/quarry tile;
   c.   Windows: UPVC/timber painted/aluminium.

2.   A client is considering replacing his heating system. System A is the standard scheme whereas system B relies on additional insulation being provided. Evaluate the alternatives and make a recommendation.

	System A	System B
Initial costs		
Boiler	160,000	175,000
Pipework and units	48,000	42,000
Insulation	12,000	32,000
Recurring costs		
Repairs (per annum)	3,000	2,800
Replacement	40,000 (every 20 years)	32,000 (every 30 years)
Overhaul	15,000 (every 5 years)	15,000 (every 10 years)
Fuel (per annum)	15,000	11,000

The expected life of each building is 60 years and the discount rate to be used is 4%.

Author's comment: For the present value of a single sum and the present value of an annuity see *'Parry's Valuation and Investment Tables'* by A.W. Davidson (*Estates Gazette*), 2002.

3.   Refer to the RICS Research Paper, vol. 4, no. 18, dated April 2003 'Whole Life Costing in Construction: A State of the Art Review' by Kishk, Al-Hajj, Pollock, Aouad, Bakis and Ming Sun.

    a.   Plot the use of WLC on a typical project against the stages within the RIBA plan of work;

    b.   Identify the characteristics of the available WLC software;

    c.   Critically review the main findings of the Report.

Source: www.rics.org/NR/rdonlyres/E4E31B2A-BC71–4C79-A73C-8280EF283EB2/0/whole_ life_costing_in_construction_20030401.pdf (accessed 2 June 2007).

Bibliography

Ashworth, A. (2004) *Cost Studies of Buildings*, 4th edition, Pearson Prentice Hall

Ashworth, A. and Hogg, K. (2000) *Added Value in Design and Construction*, Longman

Bourke, K., Shillpa Singh, V.R., Green, A., Crudgington, A. and Mootanah, D. (2005) *Achieving Whole Life Value in Infrastructure and Buildings*, BRE Bookshop

Brandon, P.S. (ed.) (1992) *Quantity Surveying Techniques New Directions*, Blackwell Scientific

Clift, M. and Bourke, K. (1999) 'Study on Whole Life Costing, Report Number CR366/98' Building Research Establishment, available online at: http://ncrisp.steel-sci.org/publications (accessed 10 April 2007)

Constructing Excellence (2004) 'Whole Life Costing, Factsheet', available online at: www.constructingexcellence. org.uk (accessed 1 April 2004)

Ferry, D.J., Brandon, P.S. and Ferry, J.D. (1999) *Cost Planning of Buildings*, 7th edition, Blackwell Science

Flanagan, R. and Norman, G. (1989) *Life Cycle Costing for Construction*, RICS

Flanagan, R., Norman, G., Meadows, J. and Robinson, G. (1989) *Life Cycle Costing Theory and Practice*, BSP Professional Books

Green, A. (2005) 'What we need is whole life value', *Building* magazine, 23 September, p. 58

HM Treasury (1992) 'Public Competition and Purchasing Unit Guidance No. 35 Life Cycle Costing', available online at: www.ogc.gov.uk (accessed 3 September 2007)

HM Treasury (2003) *The Green Book Appraisal and Evaluation in Central Government*, HM Treasury

Kirk, S. J. and Dell'Isola, A.J. (1995) *Life Cycle Costing for Design Professionals*, McGraw Hill

Lane, T. (2005) 'What we need is whole life value', *Building* magazine, 23 September, p. 58

Martin, J. (2007) 'It's always a forecast', *Journal of RICS Construction*, February, pp. 16–17

Office of Government Commerce (OGC) (2005) *Achieving Excellence in Construction Procurement Guide 07, Whole-life Costing*, OGC

Robinson, G.D. and Kosky, M. (2000) 'Financial Barriers and Recommendations to the Successful Use of Whole Life Costing in Property and Construction', CRISP, available online at: http://ncrisp.steel-sci.org/publications (accessed 10 April 2007)

Royal Institution of Chartered Surveyors (RICS) (1999) *The Surveyors' Construction Handbook – Section 2: Life Cycle Costing*, RICS

Useful websites

www.bcis.co.uk
www.bre.co.uk
www.bsria.co.uk/press/?press = 345
www.constructingexcellence.org.uk
www.hm-treasury.gov.uk
www.ncrisp.org.uk
www.ogc.gov.uk
www.wlcf.org.uk/Links.htm

Part IV
Procurement strategies

10 Organizational methods (part A)

10.1 Introduction

> The contract strategy determines the level of integration of design, construction and ongoing maintenance for a given project, and should support the main project objectives in terms of risk allocation, delivery incentivisation and so on.
>
> (OGC, 2003)

The chosen strategy influences the allocation of risk, the project management requirements, the design strategy and the employment of consultants and contractors. The contract strategy has a major impact on the timescale and the ultimate cost of the project (see Fig. 10.1).

The following four sub-sections should be considered:

1. Organizational method, for example traditional, design and build management;
2. Type of contract, for example lump sum, admeasurement and cost reimbursable/target cost;
3. Bidding procedures, for example open, selective, two-stage, negotiated, EU regulations;
4. Conditions of contract, for example would include the following:
   JCT 05 (SBC/Q), JCT 05 DB, JCT 05 MP, JCT 05 CM/A, *ICE Measurement*, 7th edition, *ICE Design & Construct*, IChemE lump sum (Red Book), IChemE reimbursable (Green Book), NEC ECC, 3rd edition, FIDIC etc.

Overview of organizational method

The RICS *Contracts in Use* surveys for building work in the UK have shown the following trends (by value):

	1991 %	1995 %	1998 %	2001 %	2004 %
Lump sum – firm BofQ	48	44	28	20	23
Lump sum – specs. and drawings	7	12	10	20	11
Lump sum – design and build	15	30	41	43	43
Target contracts	—	—	—	—	12
Remeasurement – approx. BQ	3	2	2	3	2
Management contracting	8	7	10	2	1
Construction management	19	4	8	10	1
Partnering agreements				2	7

The latest RICS *Contracts in Use* survey of building contracts in use during 2004 (RICS, 2006) showed that the decline in the use of the traditional Bill of Quantities (BofQ) identified in the 2001 survey appears to have been reversed. Likewise lump sum – spec. and drawings has declined to its 1998 level of 11% to be replaced with target contracts at 12% reflecting the use of the NEC family of contracts (now endorsed by the UK's HM Treasury's Office of Government Commerce and chosen for the 2012 London Olympics).

Design and build continues to be the most popular strategy in the building sector with 43% of the total contracts by value. The RICS survey for 2004 shows a steady increase in partnering agreements reflecting the increasing popularity of the PPC2000 contract with the Social Housing Associations.

In the UK civil engineering and infrastructure sectors, there has been a significant reduction in the use of the traditional ICE Conditions of Contract Measurement Version 7th edition. The NEC Engineering and Construction Contract family of contracts has swept all before it with most clients choosing the Activity Schedule approach (either option A – Priced Contract or increasingly option C Target Contract). This latter approach enables the sharing of risks and encourages innovation.

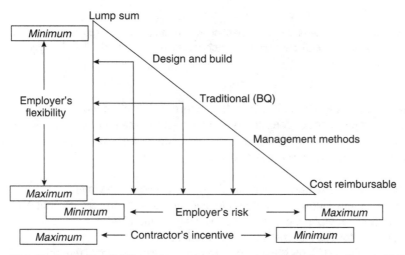

10.1 Characteristics of different types of procurement strategies (source: Barnes, NML (1983)).

Panel 10.1 Case study: Hong Kong airport and supporting rail and road infrastructure

The framework shown in Fig. 10.1 is important as it shows the various options available to clients depending on their attitude to risk.

The Hong Kong airport at Chep Lap Kok with supporting rail and road infrastructure was a true mega project; it comprised ten inter-related projects (over 200 works contracts) with four separate sponsors – Hong Kong Government, Airport Authority, Mass Transit Railway Corporation and Western Harbour Tunnel Company. The £155 billion project was completed in 1998 after six years of construction using a British–Chinese–Japanese JV – at its peak with some 35,000 workers.

In Hong Kong, the conditions of contract are normally onerous with clients wishing to transfer risk to the contractors. On the airport project the conditions of contract were extremely onerous with lump sum, fixed-price contracts adopted to the greatest extent possible to enhance certainty of final project cost. This approach is in direct contrast to the Heathrow Terminal 5 project where BAA have taken all the risk and managed the project through the use of integrated teams.

The important aspects of the Airport Core Programme (ACP) conditions of contract were as follows:

- Provision for lump sum, fixed-price form of contracts and an owner-controlled insurance programme;
- Provision for the employer, through the engineer, to make variations (additions or deletions of works) to benefit timely completion of other ACP contracts;
- Employer's ability, through the engineer, to order acceleration of work and to order contractors to recover their delays at no cost to the employer;
- Provisions for employer-referable decisions relating primarily to extensions of time and additional payment to allow a greater degree of control over the actions of the engineer on these matters;
- Stringent provisions for claim notification (which could result in the rejection of claims if not observed by contractors) so that the employer was informed of events likely to be disruptive to the programme and/or had cost implication at an early stage;
- Introduction of a tiered disputes-resolution process (mediation, adjudication and arbitration, with mediation mandatory) to help achieve significant time and cost savings when disputes arose.

Source: Lam, C.L., 1998.

10.2 Traditional method

The traditional method (see Fig. 10.2) of procurement is based on the rigid separation of design and construction. The client, usually after undertaking a feasibility study, appoints a team of consultants (led by the architect/engineer) to undertake the detailed design. The design team prepares detailed drawings, specification and often a BofQ. The tender documents are prepared and the contract awarded, usually to the contractor with the lowest bid. The contractor manages the construction using its own subcontractors together with nominated (under JCT98) or named (under JCT05) subcontractors.

The lead designer is usually in control throughout all stages of the project, from conceptual design through design development, tendering, contract administration, supervision of the works and finally to handover of the completed project.

The traditional route is readily understood but has become less popular as more clients become aware of the potential high risks carried by them if the design is not complete or the BofQ not accurate.

The traditional system

Strengths

The strengths of the traditional approach are well documented and include

- generally a high degree of certainty on the basis of the cost and specified performance before a commitment to build; however, variations and claims can make this less so;

- clear accountability and tight control at every stage; again, variations and claims can make this less so;
- competitive prices between main contractors;
- opportunity to combine best design and contracting skills in well-understood relationships;
- scope for nomination of particular specialists by client;
- flexibility in developing the design up to the contract documentation stage, and if necessary varying the construction design; however, the costs can become less certain;
- well tested, in practice and in law;
- the client is able to recover costs from the main contractor in the event that the latter fails to meet contractual obligations;
- flushes out ambiguities in the documentation prior to tender.

Weaknesses

Some of the weaknesses of the traditional approach would include

- uneasy, guarded relationship between the parties – can easily become adversarial;
- engineer/architect has no liability for the performance of other members of the design team;
- client has no right of communication or instruction with the contractor;
- overall programme may be longer than for other strategies – alternative methods allow earlier commencement on site and completion;
- no opportunity for contractor to influence design or buildability during the design process;
- split responsibilities – client is in direct contract with many different parties which can be a serious weakness in the event of major defects arising;
- does not encompass Egan's recommendations: 'fully integrated design and construction team'.

Employer

Pre-contract design
Architect/engineer/specialist contractors

Construction
Contractor/subcontractors (domestic/named/nominated)

Contracts
JCT2005 (SBC/Q)
NEC3/ ICE 7th
FIDIC Red Book

10.2 The traditional system.

Panel 10.2 Case study: traditional procurement – Bath Spa

Initially estimated at £11 million, the Bath Spa project, a public leisure amenity, in the historic town of Bath in Somerset, England, has become one of the most delayed and disrupted projects in recent times.

The project was funded via a £7.78 million grant from the Millennium Commission who specified that the council must opt for a traditional JCT contract. The aim was to open for Christmas 1999; however, in the event the contractor Mowlem did not start on site until 2000.

When the architect Grimshaw produced designs with more clarity, the costs rose to £15 million then to £22 million.

In 2003 with construction complete, Grimshaw, as contract administrator, refused to certify practical completion because it had become apparent that the RIW Toughseal paint on the walls was peeling. In September it was reported that the steam room floors were also leaking.

Grimshaw claimed that the defects were construction related, while Mowlem claimed they were design related.

In February 2005 Mowlem offered to complete the project for £26.5 million under a design and build arrangement and drop all claims against the council. This proposal was never accepted.

In April 2005 following claims that Mowlem had refused to obey an architect's instruction requiring replacement of the floors in the steam room, Mowlem were told to leave the site.

A new project manager Capita Symonds was appointed to take the scheme to completion under an arrangement in which the council would have direct control of the contractors. Capita Symonds later said they found a series of significant structural problems from leaking floors, rusty and outdated fittings and £700,000 worth of windows which had begun to delaminate.

It is anticipated that the project will be completed in 2006 with an estimated final account of £40 million.

Sources: *Building* (11 February 2005, 18 February 2005, 8 April 2005, 13 May 2005); *The Times* 24 April 2006.

If speed is a priority it is possible, however, to use accelerated traditional methods, usually through the use of two-stage tendering or negotiated tendering procedures. These enable design and construction to run more closely, securing some time saving but giving less certainty about cost.

The advantages of accelerated traditional methods are as follows:

- Two-stage tendering allows early testing of the market to establish price levels and gives early contractor involvement resulting in speed of construction;
- Negotiated tendering allows early contractor involvement for *fast tracking*, that is, beginning work on site before the design is complete;
- Negotiated tendering also gives flexibility for design development as the construction proceeds.

The possible disadvantages are as follows:

- Less certainty on price before a commitment to build;
- Competition may be diminished in negotiated tendering;
- More concentrated client involvement required to ensure efficient planning and control throughout the process.

As a post script to the traditional procurement system it is worth mentioning procurement of public works in New York city which seems out of step with the general international trend towards the use of alternative contract strategies.

New York city statutes require that contracts by state agencies for construction work be awarded to the lowest responsible bidder based on open competitive bidding. These requirements essentially mandate the traditional design-bid-build (DBB) approach to public works procurement, since competitive bidding requires a completed design, meaning that design services must be procured separately and before construction work is procured, and that the selection process cannot be based on comparative evaluation of proposals (Raved, 2003).

Anecdotal evidence from an architect who has worked in New York identified the high level of corruption prevalent in the labour unions as a possible reason for the move away from a partnering strategy back to the traditional lump-sum, fixed-price approach.

10.3   Design and build

Back in 1964 the Banwell Report remarked that 'in no other industry is the responsibility for design so far removed from the responsibility of construction.'

Today the contractor-led design and build procurement route (see Fig. 10.3) is now established as the most popular procurement route. Furthermore, it is increasingly perceived as the appropriate answer for large and complex projects, sometimes designed by signature architects.

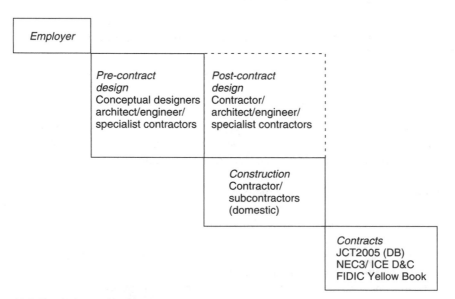

10.3 The design and build system.

The process

The design and build strategy requires the contractor to take overall responsibility for both design and construction in return for a fixed price lump sum. However, in practice, the client may appoint an independent project manager or quantity surveyor to safeguard their interests.

The client enters into a single contractual relationship with the contractor to design and construct the project in accordance with a performance specification prepared by the client. The contractor then enters into a series of separate agreements with consultants, specialist subcontractors and suppliers to deliver the project in accordance with the agreed performance specification. Since the contractor becomes solely responsible for all aspects of the project delivery process, most of the risks associated with design and construction are therefore borne by the contractor giving the client greater protection.

The client generally invites tenders based on an outline design, critical specification and workmanship standards, completion time and other key information. At the earliest stage of the construction period, the contractor completes the outstanding design development, thus generating an overlap between the design development stage and construction stage which should in theory reduce the overall duration.

In practice, there may be two separate design teams, one employed by the client to develop the client's brief and the other employed by the contractor to undertake the detailed design work. It should be noted that the former design team may not be contractually linked with the contractor though in some cases this design team may be novated to the contractor. Equally, the design team employed by the contractor is usually contractually remote from the client, and the client can only influence the output of the design through their intervention in the design-approval process. This often leads to variations and changes in the original requirements.

Contractor's expertise

The design and build approach allows the contractor's design and construction team to consider, at the earliest conceptual stage, site-specific construction issues which a consultant working in isolation is not normally equipped to deal with. For example, on a large marine project the team will be able to establish the following criteria: if the site is suitable for the use of large cranes; whether heavy floating barges can be used in a tidal location; how materials will be transported to the construction locations; whether there are suitable areas close to the site for setting up a precasting or pre-assembling yard; what skills are characteristic of the local labour force and how the local weather during the construction period will affect the construction methods.

The most economic type of structure and the most suitable method of construction will depend on the answers to the above questions, together with the contractor's specific expertise and the availability of construction equipment. It is at this stage when the combined team has at its disposal all the relevant facts and techniques that increased productivity may be considered – thus reducing the overall cost to the client.

Flexibility

The design and build route is extremely flexible and many different versions have emerged over the past decades. The major difference between them is the amount of design input by the employer's designers and the contractor's designers; for example; the client's contribution to the design may vary from 5% to 75%.

The amount of tender documentation provided by the client (known in the JCT Standard Form of Building Contract with Contractor's Design 2005 as the *Employer's Requirements*) can vary from little more than a written brief to a fully worked out scheme. The greater the priority

---

Panel 10.3 Case study: design and build – Cardiff Millennium Stadium

The main reason Laing lost so much money (£31 million) on the £99 million Millennium Stadium in Cardiff was that it guaranteed a maximum price on the basis of sketchy designs which were still undergoing change.

The original design had masts raking out at 45 degrees at the four corners of the stadium, but a row between the stadium operator and its neighbours Cardiff Rugby Football Club led to these being revamped to two vertical and two raking masts.

The load calculations for the £480,000–member roof had to be redone. The problem was that the design was still developing as Laing fixed its price and was still developing as the roofing contractor hit the site.

Cardiff Ruby Football Club also refused to allow Laing's tower cranes to swing over its air space, cutting off access to one side of the ground.

Claims came in from subcontractors, and under the contract Laing carried the risk and were left with the bill.

Shortly after the stadium opened Laing were required to install 74 giant metal props in order to make the cantilevered stands safe for a 'new year eve' rock concert. A legal expert close to the project said that the contract documents were an absolute mess and that the situation, on liability, could be hard to resolve.

After this project the Laing Group was reorganized with the construction arm being sold to concrete specialist O'Rourke for £1 million.

Source: *Building*, 17 September and 17 December 1999.

---

the client gives to design the larger the amount of information tends to be included in the tender documents. If a client's priorities are economy and speed then less design information will be included, leaving more scope to the contractor.

Three main categories of design and build approach can be identified:

1. *Direct:* designer/contractor appointed after some appraisal but not competition;
2. *Competitive:* conceptual design prepared by consultants, several contractors offer designs in competition;
3. *Develop and construct:* client's designers complete design to partial stage before asking contractors to complete and guarantee the design in competitive tender either with their own or using the client's designers (novation).

Responsibility for design

There are two standards of care which are relevant to the design and build strategy:

Reasonable skill and care

A duty imposed on a professional consultant who provides advice or a service. It is effectively a matter for professional judgement whether in providing that advice or service the consultant has exercised all the skill and care that can reasonably be expected. Only if professional negligence can be proven is the consultant liable for failure of the end product.

Fitness for purpose

A statutory requirement under the Sale of Goods Act and an implied term in any contract for the design and supply of a finished product. If the product is proved unfit for the intended purpose for which it was supplied, then irrespective of whether the provider of the product has been negligent or not, there would be a liability for any failure of that product to perform.

The risk of the design meeting the client's requirement would be spread differently depending on the contract strategy adopted.

Under the traditional, management approach:

- The consultant(s) who design the project would not be liable for the performance of the completed project unless it could be proved that the consultant had failed to exercise the level of skill and care reasonably to be expected. Before recovering damages, the sponsor must prove actual professional negligence by the consultant, often a substantial hurdle to overcome in any litigation.
- The contractor(s) who only builds the project using the design provided by the consultant(s) would be liable for fitness for purpose of the materials and components that were provided within the project. The contractor(s) would not be responsible for the performance of the works as a whole.

Under a design and build strategy:

- The contractor who designs and also builds a project would be liable not only for the fitness for purpose of the materials and components, but also for the completed project as a whole. Not surprisingly, it is unlikely that a contractor would be able to obtain professional indemnity insurance while taking on such a design obligation.

In the case of *Co-operative Insurance Society* v. *Henry Boot (Scotland) Ltd TCC 1 July 2002* the contractor was deemed to have effectively audited and adopted as its own, the design work of others. The fact that the piling, the offending elements of works, had been designed by

---

Panel 10.4 Case study: design and build – Hammerson

For large schemes, generally of more than £20 million, Hammerson use a two-stage bidding process. They invite a few major contractors to work with their designers to develop the scheme and select one of them based on issues such as preliminary costs and overheads. When the design is 80% complete, they will novate the design team to the contractor. The contract then becomes a design and build contract or one with a target price guaranteed.

Source: '50 top clients a building directory', *Building*, Supplement, February 2003.

consulting engineers did not protect the contractor who possibly unwittingly took on an obligation to complete the design. Following this case it is noted that the JCT Design and Build Contract (DB2005) now provides that the contractor is not required to check the design in the employer's requirements and will not be responsible for any adequacy in it.

The design and build system

Strengths

- Early completion is possible because of early commencement date and overlapping activities;
- Single-point responsibility for total design and construct process after the selection stage – one back-side to kick if anything goes wrong!
- The client can demand the quality and the performance specified;
- Price certainty is obtained before construction starts provided the client's requirements are adequately specified and changes are not introduced;
- Varying amount of client design input can be allowed;
- Input by the contractor can lead to more economic design;
- It is less adversarial than the traditional approach;
- Tender competition on alternative design solutions is possible;
- Direct lines of communication between subcontractor and designer is possible;
- Due to early collaboration between design and construction disciplines variations are reduced.

Weaknesses

- Competing schemes may not meet client's requirements unless specified in detail before the bidding begins;
- The cheapest building in terms of whole-life costs may not be produced;
- The client loses control over quality in design;
- Client's clearly defined brief required at commencement;
- Changes after commencement are expensive;
- Analysis of tenders may be subjective.

The employer's agent

Many cost consultants are now employed as the employer's agent (EA) under the design and build procurement route. The role taken by the EA must be clearly defined by the client and can vary from simple contract administration to a full project management service. Beyond specific contractual duties other areas in which the EA can contribute include the following:

- drafting the project execution plan;
- managing the briefing and scope-definition process;
- appointing design consultants, on a basis that facilitates the effective transfer of design responsibility;
- managing clients and third-party liaison;
- preparing employer's requirements and other tender documentation;
- implementing change control;
- supporting use of warranties to meet the requirements of third parties.

For a full discussion on the role of the EA, see Simon Rawlinson's detailed article in *Building* magazine (9 February 2007), pp. 58–61.

Panel 10.5 Case study: design and build – Wembley Stadium

When the contract to build the Wembley Stadium was first put out to tender in late 1999 on a fixed-price lump-sum basis Sir Robert McAlpine declined to bid, warning that 'The sums of money need to adequately cover (the risk of the builders running over budget) would be expressed in tens of millions of pounds.'

In February 2000, Multiplex, a major Australian contractor submitted a bid in joint venture with UK contractor Bovis and were appointed as the preferred contractor. Bovis refused to accept a price reduction below £339 million and on 30 August the client WNSL terminated the joint venture. Two days later Multiplex made a solo offer to build Wembley in 39 months for £326.5 million which was accepted ten days later by WNSL 'subject to board approval'. In the negotiations that followed, after funding was secured, the figure was increased by £120 million to a guaranteed fixed price of £445 million.

This 40-month design and build project had the advantage of a late start with a well-developed design and few changes were envisaged. Any savings made by the contractor would be retained by them; likewise they would carry all risks. Multiplex told their shareholders that they was confident of securing a profit on the project.

Wembley's great technical challenge was the structural design, construction and erection of its signature arch. Multiplex awarded steelwork specialist Cleveland Bridge an 81-week, £60 million lump-sum fixed-price subcontract to fabricate, supply, deliver and erect the arch and roof.

However, in 2004 the site was severely disrupted between March and August as a legal dispute developed between Cleveland Bridge and Multiplex. Cleveland Bridge were subsequently ordered off the site and replaced by Dutch steel subcontractor Hollandia working on a 'cost-plus' basis.

In a claim against Multiplex for non-payment, filed at the Technology & Construction Court, Cleveland Bridge alleged that 'by the Spring of 2003 there were serious problems arising from late and incomplete design by the civil and structural engineers, Mott Stadium Consortium (which had been novated to Multiplex), and delays in providing design information. The design changes and late information caused substantial cost increases, and delays and disruption to the subcontract works.'

Though Cleveland Bridge and Multiplex agreed on a plan for accelerating work, plus compensation for the resulting change in the subcontract terms, a legal row broke out, over alleged non-payment and contract breaches.

In May 2005 Multiplex announced losses of £45 million; their shares crashed and were suspended on the Australian stock market. These losses for Multiplex did not take into account the costs associated with claims from Cleveland Bridge (High Court writ claims £20 million) or from the other subcontractors claiming a 15-week delay and associated disruption (estimated at £20 million).

An announcement said the extent of the losses was dependent on five factors: the ability to successfully recover claims against third parties, the ability to meet the construction programme, costs associated with the project's steel work, the cost of the project preliminaries and, as required, acceleration and the weather. To date the project has so far generated seven court judgements.

After extensive delays Wembley Stadium was finally handed over to the client in March 2007. It was reported that Multiplex made a loss of £147 million on the project and the Australian family company was sold in June 2007 to a Canadian firm.

Source: Mylius, A. (2005) Supply Management, *Building* (31 May 2002, 11 October 2002, 29 October 2004, 3 June 2005, 4 November 2005, 15 June 2007); Conn, D. (2006); *Guardian*, 8 March.

10.4 Turnkey

The turnkey contract has been adopted for many years on major multidisciplinary construction projects, particularly in the process-plant sector, both in the UK and overseas. Under turnkey projects the entire process of design, specification, construction and commissioning is carried out by contracting organizations often in either joint venture or consortia. Sometimes the client may wish the contractor to finance, operate and maintain the facility.

The client will normally issue a brief based on a performance specification together with outline drawings indicating a preferred layout. The contractor's lump-sum bids are evaluated first on technical and performance basis and second on a financial basis for capital expenditure and running costs (using the discounted cash-flow technique). The advantages of the turnkey approach include single-source responsibility relieving the client from the responsibilities for equipment and performance, a fast-track approach with design and construction overlapping and a lump-sum price. The disadvantages of the turnkey approach include lack of client control and participation, significantly higher overall cost than the traditional approach, and very limited flexibility to incorporate changes.

Panel 10.6 Case study: turnkey contract – Innogy Holdings

Innogy uses several different procurement routes depending on the project. Large power stations, such as the Staythorpe plant, which is currently on hold, are planned to be let under a turnkey contract.

For smaller power projects, the company appoints consultants that let the contract, most likely on a competitive tender basis. For such work as the servicing of boilers, Innogy has power-engineering alliances.

Office projects are generally design and build contracts, and large refurbishment schemes are competitively tendered.

Source: '50 top clients a building directory', *Building*, Supplement, February 2003.

Panel 10.7 Case study: design and build to turnkey – Arsenal's Emirates Stadium, London

In contrast to Wembley Stadium, Arsenal's new 60,000-seater, £275 million Emirates Stadium opened two weeks ahead of programme on 22 July 2006, despite incorporating £35 million of extra works during its construction.

The stadium was only part of a £390 million development which involved the relocation of all the businesses on the new site, building a state-of-the-art recycling centre to re-house Islington Council's refuse disposal service and large-scale regeneration in the area involving providing day care and health centres and 2,000 new homes.

The construction contract was the key project document that enabled the club to raise the debt finance to construct the stadium on a non-recourse basis. As with Wembley, the construction contract started life as a JCT standard form with contractor design. As the financing developed, amendments were negotiated to convert it to a lump-sum fixed-price, fixed-date 'turnkey' contract where most of the construction and programme risk lay with Sir Robert McAlpine, the contractor for the job.

McAlpine was prepared to accept single-point responsibility for design and construction on the basis that, on the conversion of the contract from reimbursable to fixed price, the professional team was novated to it. The professional appointments had been signed in advance with the novation in mind to provide the contractor with recourse against the professionals in event of a claim under the construction contract.

The Emirates stadium was voted *Building* magazine's 'Project of the Year' in April 2007.

Sources: *Building*, 9 June 2006, 23 June 2006, 20 April 2007 (supplement).

10.5   Joint ventures

Joint venture arrangements are an attractive means of cooperation between organizations. They are used in many industries and countries for temporary or selective cooperation. Most typically they are temporary arrangements for the purpose of carrying out one project – usually a major project. A joint venture can be defined as number of firms collaborating on a project or a number of projects with a view to sharing the profits, each firm being paid on the basis of its agreed contribution in kind or in financial terms.

In general, joint ventures are not the easiest forms of organizations to manage and operate. Therefore, there must be compelling reasons why parties to a construction contract resort to a formation of a joint venture in contrast to the conventional contractor/subcontractor relationship; these could include the following:

Pooling of resources and expertise

The author worked as senior quantity surveyor (QS) for Gammon Kier Lilley (GKL) joint venture on the North Nathan Road project on the first stage of the Hong Kong MTR. The Hong Kong firm Gammon provided all local Chinese supervisory labour up to general foreman level throughout the project; Kier were responsible for constructing the three underground stations (Waterloo, renamed Yau Ma Tei; Argyle, renamed Mong Kok and Prince Edward stations), with Scottish tunnelling specialist Lilley responsible for all the tunnelling between the stations.

Sharing of risks

The main risks on construction projects have been identified by Perry and Hayes (1985) and include the following:

* Physical – ground conditions/loss or damage by flood or fire;
* Environmental – public inquiry/pollution;
* Design – new technology/incomplete design/temporary works;
* Logistics – transportation/access;
* Financial – cash flow/exchange rates/adequacy of insurance;
* Legal – liability for others, local law, conditions of contract;
* Political – war/revolution/imported licences;
* Construction – feasibility of methods, rates of production, industrial relations/safety;
* Operational – fluctuations in market demand/maintenance/fitness for purpose.

In the event, on the Hong Kong, North Natham Road projects the major problems concerned ground conditions – isolated rock boulders were encountered when constructing the secant-piled station walls. The large diameter piling-auguring equipment, which had been mobilized from the UK, proved to be inadequate with the only solution being the old Chinese technology of hand-dug caissons. However, before these could be constructed, four grout curtains were required to be built on the outside and inside of the station walls, in order to stabilize the ground and keep out the water. This massive additional expenditure and subsequent delay and acceleration costs became the subject of a multimillion pound claim.

Entry to foreign markets

An overseas contractor can enter a local market with some degree of comfort, by establishing a joint venture with an established contractor. Indeed, many countries have enacted foreign investment controls, which require foreign companies to enter into arrangements with one or more local companies with a view to imparting their expertise to the local firms in order to decrease the country's dependence on foreign expertise in the future.

Access to technological improvements

Few companies can maintain cutting-edge knowledge in all their critical activities. Corporate partnerships can be used to gain access to new competencies and technologies.

Main types of joint ventures

Joint ventures can take a variety of forms with varying degrees of complexity. The three main types can be classified as follows:

A simple contractual relationship

Contractual joint ventures are entirely governed by contract and do not involve the establishment of a legal separate identity. One or more contracts can define the manner in which the joint ventures are to be conducted and the respective duties and responsibilities of the parties.

In a contractual joint venture, unless they agree to the contrary, each of the parties will be liable for its own acts and omissions. In addition, contractual joint ventures are transparent for tax purposes, leaving each party to arrange its own tax affairs without interference from the other.

A partnership agreement

Unlike contractual joint ventures, the establishment of a partnership provided the joint-venture partners with an independent vehicle capable of trading under its own name. The establishment and operation of a partnership under English law is governed by the Partnership Act 1890 and by case law.

In practice the partnership structure is employed relatively infrequently in commercial joint ventures. The principal reason for this is the perceived exposures arising from the fact that each partner is personally liable for the full amount of the partnership's liability to third parties.

A limited liability company

The joint venture company (JVC) is the most common form of organizational structure where the parties wish to establish and operate a jointly owned business. The JVC, unlike a partnership, will have a distinct legal entity with separate interests from its members. A JVC is thus able, under its own name, to sue and be sued and enter into contracts with third parties.

The company's objects and powers are set out in its Memorandum of Association. The regulations governing its conduct are set out in the Articles of Agreement. In addition, JVCs also complete the Shareholders' Agreement in which the parties would normally set out the manner of establishing, funding and operating the JVC.

The formation and operation of a JVC is complex compared to a partnership. These complexities result from the fact that the constitution, operation and rights of shareholders, are prescribed by statute and case law.

The advantages of the JVC are generally perceived as threefold:

- The maximum liability of shareholders in respect of the JVC is the amount paid up, or agreed to be paid up;
- The JVC structure operates within a familiar body of law and practice;
- JVCs are better placed to raise external financing.

10.6 Consortium

A consortium is similar to a joint venture, that is, an arrangement between several firms, but in this case each contributes an equity stake in the form of risk capital or payment in kind in order to qualify as a member. Remuneration of consortium members may be calculated as a share of net profits of the consortium. Consortia aim to work together as a team solving problems, reducing costs, lessening risks and addressing quality issues in a shorter period than with traditional procurement methods.

Consortium is particularly attractive on PFI projects where construction and operating companies can be brought together under one organization, combining their skills and sharing the risks.

Both joint ventures and consortia invariably set up Special Purpose Vehicles (SPVs). Under this arrangement a formal accounting and contractual arrangement is set up separate from the individual accounts of the firms involved.

However, new research from the University of Reading (Gruneberg and Hughes, 2006) indicates that in practice firms form a consortium only where no alternative structure would be feasible. Unlike partnering, it does not seem that a consortium offers a more stable and long-term approach to client relations, and members do not necessarily go on to work in the same consortium again.

Furthermore, working in consortium, participants may carry extra risk in the form of liability for the actions of their partners. The University of Reading research identifies that consortium is not a way to deliver best practice but is largely a marketing device to win business.

10.7 Partnering

Japanese contractors have an international reputation for achieving quality, certainty of outcome and completion on time. This great success is based on long-term relationships based on trust and a sense of brotherhood. Customers, general contractors, suppliers, specialist subcontractors and subcontractors have worked together in tightly knit families of firms for decades. Under this arrangement the general contractors set tough standards but take responsibility for the well-being of their subcontractors (Bennett, 1991).

In the last decade, one of the most significant developments in procurement in the UK has been the use of partnering. Essentially partnering promotes improved performance through collaborative business relationships based on best value rather than lowest cost. Contract awards are still subject to rigorous competition but judged on predetermined combinations of quality and cost. The development of openness and trust is in contrast to the confrontational nature that has increasingly characterized much of the construction industry over recent decades.

Professor John Bennett and his research team at the University of Reading identified partnering best practice within *Trusting the Team* (Bennett and Jayes, 1995) and *The Seven Pillars of Partnering: A Guide to Second Generation Partnering* (Bennett and Jayes, 1995).

> Partnering is a management approach used by two or more organisations to achieve specific business objectives by maximising the effectiveness of each participant's resources. It requires that the parties work together in open and trusting relationship based on mutual objectives, an agreed method of problem resolution and an active search for continuous measurable improvements.
>
> (Bennett and Jayes, 1995)

Contracts have been commonly formed consisting of a partnering arrangement used in conjunction with a suitably amended standard form of contract. The partnering agreement generally defines the provisions of the arrangement, such as attitude, partnering performance, allocation of risk and incentives.

In recent years forms of contract have been drafted that incorporate partnering principles and clauses. These include the following:

- *The NEC ECC Secondary Option X12*
  This is an agreement between the client and contractor and includes partnering-type obligations such as the need to work together to achieve the client's objectives and the provision of incentives.
- *PPC2000 Partnering Agreement*
  This contract has been specifically drafted for partnering projects using the Construction Industry Council Guide. It is a contract which integrates the full partnering team, the procurement process and procedures for running the project. The procedures are very prescriptive and the timescales in some cases very short.
- *The ICE Partnering Addendum*
  This is an addendum which can be used with the traditional *ICE Conditions of Contract*, 7th edition, Measurement Version.
- *JCT Partnering Charter (non-binding) (PC/N)*
  This enables parties who do not wish to enter a legally binding agreement to create a collaborative working environment.
- *JCT Framework Agreement*
  This framework agreement can be used with most standard forms of construction and engineering contracts and with subcontracts.
- *JCT Constructing Excellence Contract (CE)*
  This new partnering contract has been endorsed by the Local Government Association and fulfils the attributes of *Achieving Excellence in Construction*.

Partnering is not a soft option because it requires considerable effort to set up and hard work to maintain, but most people who have experienced partnering have found it both satisfying and enjoyable.

Panel 10.8 Case study: the Highways Agency

The Highways Agency is an executive agency of the Ministry of Transport with the responsibility for managing, maintainting and improving the network of trunk roads and motorways in England. They have made considerable progress since their first strategy document was launched in 1997. The new strategy, embracing the *Rethinking Construction* philosophy will build on the successful initiatives and pilots which have been implemented since then and which include the following:

- Design and build contracts used for all major improvement schemes. The ECC contract now used for nearly all other works contracts.
- The first 'early design and build' contract awarded on the A500 Stoke pathfinder project.
- Project-partnering arrangements applied to major contracts.
- Improved dispute resolution procedures and reduced numbers of disputes.
- Most new contracts awarded on the basis of quality and price.
- New forms of maintenance contract implemented including the first single-point supply Managing Agent Contractor (MAC) contract.
- Framework contracts introduced for regional works projects and design services.
- New payment mechanisms linked to the level of service to road users introduced into new private finance DBFO contracts.
- A construction management framework successfully piloted to assess the benefits of an integrated supply chain and the measurement of actual costs.
- A new supplier database implemented together with improvements to supplier performance reporting procedures.

Source: www.highways.gov.uk/business/933.aspx (accessed 10 January 2006).

Benefits of partnering include

- increased customer satisfaction;
- better value for client;
- recognition and protection of profit margin for contractors and suppliers;
- staff development and satisfaction;
- creation of an environment that encourages innovation and technical development;
- better understanding between partners and driving down of real costs;
- design integration with specialists in the supply chain;
- improved buildability through early involvement of the contractors;
- elimination of duplication;
- better predictability of time and cost;
- shorter overall delivery period;
- improved quality and safety;
- stability which provides more confidence for better planning and investment in staff and resources.

Early partnering arrangements tended to be on a one-to-one basis between combinations of client, main contractor and professional services providers. However, multiple partnering is now common with a number of parties bound under the same agreement, dependent on and cooperating with each other for overall success.

Although normally client led, there are many examples of contractors and suppliers demonstrating the advantages to their clients, who have subsequently chosen this method of procurement. To gain maximum benefit it is essential to extend the process through the supply chain in order to harness the specialist expertise of subcontractors, material suppliers and manufacturers.

Much of the information in the section on 'Partnering' (Section 10.7) was taken from the *Constructing Excellence Partnering* fact sheet (www.constructingexcellence.org.uk). Partnering seems the way forward for major organizations and substantial benefits are being realized. At the end of 2004, four groups of companies entered into framework agreements with Yorkshire Water to undertake improvements to the region's water and sewerage systems over a five-year period; Severn Trent Water have had a similar arrangement in place for several years. In 2005 The National Grid signed a £1.6 billion deal for gas mains replacement using only four partners to carry out the work.

The Welsh Water Alliance is a strategic partnering team – formed between Dwr Cymru Welsh Water, United Utilities (contracted to operate Welsh Water's assets), six strategic design/ construction partners, two cost managers, a partnering facilitator and a supply chain advisor – to deliver around 60% of Welsh Water's capital investment programme during the period 2000–2005.

For an in-depth review of project-specific alliancing based on best practice, readers are recommended to consult the European Construction Institute's excellent publication: *Partnering in Europe: Incentive-based Alliancing for Projects* by Bob Scott (Scott, 2001).

Panel 10.9 Case study: partnering Staffordshire County Council/Birse Construction Ltd

This is one of the first projects the author came across which made him appreciate the real significance of the partnering approach.

Staffordshire County Council was one of the first clients to implement partnering as an addition to a contract under the ICE 5th Conditions of Contract on the £10.1 million Tunstall Western Bypass – Phase II with a planned construction period of 15 months.

This project had all the ingredients of a problem contract – a canal, a railway crossing, contaminated ground, underpass under a busy trunk road, tight site and numerous structures.

The bulk earthworks for the scheme presented the greatest difficulties and risks. Because of the heavy industrial use of the site over the previous century, the whole site was deemed to be contaminated.

In the event, the engineer and contractor, in conjunction with the Environment Agency, worked to maximize the amount of earthworks for reuse as acceptable fill. Mark McCappin, the Resident Engineer, admitted that this one issue could have cost the client an extra £6 million, and resulted in a six-months overrun, if it had been administered in the usual adversarial manner.

In the event, the project was completed within the 67-week contract period and the final account settled within the budget.

Source: McCappin, 1996.

Panel 10.10 Case study: Birmingham Construction partnership

Against the backdrop of *Rethinking Construction* and *Achieving Excellence in Construction* and government reports into improving construction provision and procurement in the UK, Birmingham City Council founded the two-tier supply chain Birmingham Construction Partnership (BCP). This arrangement created a unique collaboration of three contractors: Wates Group, Thomas Vale and GF Tomlinson, which together form the first chain of the supply chain. The second tier comprises 61 companies from whom equipment and services are sourced. The supply chain was tasked with delivering every project in the city with a budget over £1 million over five years from 2004–2009 under a £500 million capital building programme.

The contractor is involved from the very start, at the planning and costing stage, working with the customer, the design team, subcontractors and key suppliers on the development, specification, buildability and delivery of new-build and refurbishment projects.

The contractors organize and manage the supply chain to deliver all projects on time and on budget across all the Council's services including education and social care.

Latest performance indicators show that in 2004–2005, the partnership scored above national industry averages for 2004. Key figures include

- 92% of projects delivered with zero or minimal defects;
- 61.8% of projects delivered within 5% of target cost (national average 38%);
- 62.3% of projects delivered within 5% of target time (national average 60%);
- 77.8% of projects delivered under partnering principles, compared to the government's target of 20%.

BCP has featured on www.ogc.gov.uk, the website of the Office of Government Commerce – which works to improve procurement in government and the public sector – as 'the first construction collaboration of its kind in the UK'. Its work with the Council's Housing Department on improving Council homes was selected as a best-practice case study by the West Midlands Centre of Excellence at www.wmcoe.gov.uk

It has just been hailed as an example of best practice in local authority construction in a landmark report Transforming Local Government Construction: The Power of Framework Agreements, which was launched by the Local Government Task Force in May. The report, which highlights the benefits and efficiencies that councils and contractors can achieve through partnering and collaboration, can be seen at www. constructingexcellence.org.uk

Source: www.birmingham.gov.uk (accessed 4 March 2007).

Two concepts lie at the heart of partnering and alliancing. One is the idea of *gainshare* and *painshare*. If the approach is to work, an inviting prospect of additional profits through gainshare must be put in place, with a corresponding drop in profits through painshare (usually capped for the contractor).

The second concept is that of performance assessment. The success of projects within an alliancing programme can only be gauged if measures of performance are developed and implemented (MPA seminar, 2002).

Panel 10.11 Case study: partnering – health clubs

A programme of constructing health clubs consisted of the roll-out of a large number of 20,000 sq ft complexes around the UK, with the intention of realizing incremental improvement year on year. Each cost between £1 million and £1.5 million.

In year one, a core team of industry professionals and constructors was selected, and a generic brief developed. In year two, the product was developed, formal partnering arrangements were introduced and a collaborative environment developed. In year three, the team concentrated on improving the product, with a shift of emphasis from cost to the product itself. That enabled them to increase standardization and shorten the construction periods and the process for approvals. A project-debriefing process was also set up to build in the lessons learned.

Among the achievements were

- a 25% reduction in capital cost
- increased quality
- a reduction in construction time and time to market
- approaching 100% predictability
- approaching zero defects at handover
- zero reportable accidents throughout the programme.

Source: Major Projects Association (2002).

On the international stage, successful applications of partnering in construction have been reported in the US and Australia, while in Japan partnering is considered the normal way of working. In Hong Kong, the Mass Transit Railway Corporation (MTRC) pioneered the partnering approach on the sixth stage of their metro system.

A case study examination on the Yau Tong Station (Contract 604), one of the 13 civil projects on the Tseng Kwan O Extension identified two key partnering tools. First, the monthly partnering review meeting at which the partnering status was assessed by considering 13 attributes. Second, the Incentivization Agreement which was signed mid-way through the contract; this introduced a target-cost arrangement for dealing with shared risks using a pain/gain formula (Bayliss *et al.*, 2004). However, UK construction lawyer Roger Knowles identifies that somewhere in the region of 80% of construction work on new-build projects is undertaken by subcontractors and they may not be included in the partnering process. He further identifies that some main contractors who profess to embrace partnering still use one-sided conditions of contract (Knowles, 2005).

Rudi Klein, Chief Executive of the Specialist Engineering Contractors Group, also confirms that one of his members received a heavily amended NEC ECC Construction Subcontracts, while part of a supply chain on an *NHS Procure 21* project, in effect, mirroring the old traditional approach of *dumping risk* on to the subcontractors (Klein, 2004).

Roger Knowles (Knowles, 2005) considers that in the private and public sector the jury is still out as to whether partnering brings the benefits that are claimed. Anecdotal evidence from several senior professionals within the industry tends to confirm this viewpoint.

10.8   Conclusion

This chapter has identified the benefits and weaknesses of the traditional design and build approaches. The chapter also discusses the advantages of joint ventures for large projects, and how and why partnering has been identified as the way forward for many clients.

Critical to this choice of procurement route is the allocation of risk to the client. Most clients are averse to risks and require that projects are completed on time and within budget. It is noted that in February 2007, the Irish government began replacing all existing public-works contracts with a new suite of contracts aimed at providing fixed-price lump-sum contracts. These new contracts are intended to give price certainty and eliminate the potential for claims for any extras and overruns that go beyond the original budget.

The contracts include severe terms/penalties which preclude the contractor's common law rights and introduce time limitations on the submissions of claims and programme contingencies (O'Sullivan, 2007). This approach seems similar to that adopted by the Hong Kong International Airport (refer Panel 10.1) project against which there were substantial claims.

The traditional approach with both the client and the contractor attempting to exploit each other led to the introduction of partnering and the development of long-term alliances. However, it seems that many clients in the public and private sector are not yet convinced that this approach leads to the holy grail of *best value*. Significantly, BAA's new owner and the Spanish contractor, Ferrovial has indicated a shift away from the innovative T5 construction management approach (see Chapter 18). It is now expected that most future BAA contracts will be let on design and build basis (Macdonald and Richardson, 2008).

10.9   Questions

1. Identify the key issues which might persuade a client to use the traditional procurement route.
2. Critically review the quantity surveyor's role under the design and build procurement method (refer Kennedy, S. and Akintoye, A.'s RICS COBRA 1995 Research Conference).

Source: www.rics.org/Builtenvironment/Quantitysurveying/quantity_surveyors_role_in_design_19950901.html (accessed 20 March 2007).

Bibliography

Banwell, H. (1964) *The Placing and Management of Contracts for Building and Civil Engineering Contracts*, HMSO
Barnes, N.M.L. (1983) 'Advanced Construction Management Techniques', Continuing Engineering Studies Seminar, Civil Engineering Department, The University of Texas, Austin, Texas, 2–3 November
Bayliss, R., Cheung, S.O., Suen, H.C.H. and Wong, S.P. (2004) 'Effective partnering tools in construction: A case study on MTRC TKE Contract 604 in Hong Kong', *International Journal of Project Management*, vol. 22, no. 3, April, pp. 253–263
Bennett, J. (1991) *International Construction Project Management: General Theory and Practice*, Butterworth Heinemann
Bennett, J. and Jayes, M. (1995) *The Seven Pillars of Partnering: A Guide to Second Generation Partnering*, Thomas Telford
Boonham, S. and Nisbet, M. (2004) 'Achieving Best Value on Bullring: A Case Study', Gariner & Theobold publication, RICS COBRA 2004 Construction Research Conference, September, Leeds
Gruneberg, S. and Hughes, W. (2006) 'Understanding construction consortia: Theory, practice and opinions', RICS research paper series, vol. 6, no. 1
HM Treasury Central Unit on Purchasing (CUP) (1992) *Guidance Note No. 36 Contract Strategy Selection for Major Projects* (since withdrawn) HM Treasury

HM Treasury Central Unit on Purchasing (CUP) (1995) *Guidance Note No. 51 Introduction to the EC Procurement Rules*, HM Treasury, available online at: http://archive.treasury.gov.uk/pub/html/docs/ cup/cup51.pdf

Janssens, D.E.L. (1991) *Design-Build Explained*, Macmillan

Klein, R. (2004) 'You Are the Weakest Link', *NEC Newsletter*, Issue No. 30, August, available on at: www.newengineeringcontract.com/newsletter/article (accessed 3 September 2007)

Knowles, R. (2005) 'A New Dawn for Partnering', available online at: www.jrknowles.com/legal articles (accessed 8 August 2005)

Lam, B.C.L. (1998) 'Management and procurement of Hong Kong Airport Core Programme', *Proc. Instn Civ Engrs, Civ. Engng*, Hong Kong International Airport, part 1: airport, vol. 126, no. 1, pp. 5–14

McCappin, M. (1996) Staffordshire County Council/Birse Construction Ltd, 'Partnering on Tunstall Western Bypass – Phase 2', Wulfrunian Lecture 1996, Partnering in Civil Engineering, Institution of Civil Engineering Surveyors

Macdonald, S. and Richardson, S. (2008) 'BAA to axe 200 staff from its projects division', *Building magazine*, 11, January, p. 9

Major Projects Association (2002) 'Alliance/Partnering', summary of seminar held at Templeton College, Oxford, 17 January

Minogue, A. (2004) 'Seduced by simplicity', *Building* magazine, 15 October, p. 55

Mylius, A. (2005) 'A game of two halves', *Supply Management*, 6 October, accessed via www.recruiter.co.uk/ Articles/330375 (accessed 14 January 2008)

Office of Government Commerce (OGC) (2003) *Achieving Excellence in Construction, Procurement Guide 06 Procurement and Contract Strategies*, OGC

O'Sullivan, G. (2007) 'All change in Eire', September/October, *The Journal RICS Construction*, pp. 18–21

Perry, J.G. and Hayes, R.W. (1985) 'Construction projects – know the risk', *Chartered Mechanical Engineer*, February, pp. 42–45

Raved, J.E. (Chair) (2003) 'Report on Alternative Methods of Public Works Procurement', Association of the Bar of the City of New York, Committee of Construction Law

Rawlinson, S. (2007) 'Procurement: Employer's agents', *Building* magazine, 9 February 2007, pp. 58–61

RICS (2006) *Contracts in Use: A Survey of Building Contracts in Use during* 2004, RICS

Scott, B. (2001) *Partnering in Europe: Incentive-based Alliancing for Projects*, Thomas Telford

11 Organizational methods (part B)

11.1 Introduction: management methods

The traditional system is often too rigid and has been used in inappropriate circumstances, for example, where the design is not complete. Projects have become more complex and demanding in terms of technical, legal and financial, speed, size and logistics aspects. Under complex circumstances the management strategies may offer an alternative approach. In such cases, the contractor/construction manager offers the client a consultant service, based on a fee, for coordinating, planning, controlling and managing the design and construction (see Fig. 11.1).

These approaches ensure that the contractor (or construction management consultant) is part of the client's team from the outset, and similar to the other consultants is paid a fee, ensuring that the maximum construction experience is fed into the design.

The management approach can offer a viable and flexible relationship where:

- the project is large and/or complex;
- a fast-track procurement system is required, need for an early start and early completion of the project in the situation;
- the client wishes to select the designer separately (e.g. where a competition is used);
- the work is not sufficiently defined prior to construction (enables the detailed design and technology to be developed in tandem);
- the project is organizationally complex;
- flexibility is required throughout the project;
- a less adversarial approach is preferred;
- a choice of competitive tendering for each element is preferred;
- a target price (based on a cost plan) rather than a fixed lump sum is accepted.

Contractor Bovis introduced the Bovis System into the UK in the late-1920s. Under this system the builder was paid a fixed fee to cover overheads and profit with the client receiving any savings made during the construction instead of the contractor. In 1927 Bovis signed the first contract with retailer Marks & Spencer, a prime cost fixed-fee contract, which awarded the provision of a bonus to Bovis if the actual cost was lower than the estimated cost. The marriage between client and contractor proved an outstanding success with well over 1,000, Marks & Spencer projects completed by the year 2000 (Cooper, 2000).

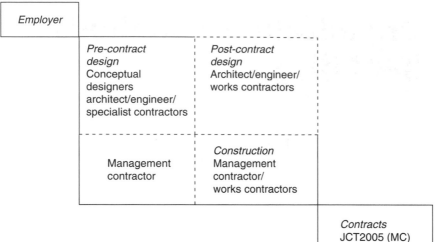

11.1 The management contracting system.

11.2 Management contracting

Following the success of the Bovis System, management contracting was a logical development for the Bovis Group in the 1970s. This approach offered the same unique relationship between client and contractor, with a negotiated fee, but included competitive tendering for subcontracting packages. Bovis were the pioneers of this approach in the UK, and their successful management contracting projects in the 1970s included John Players & Sons HQ in Nottingham, Wiggins Tea HQ in Basingstoke, Norman Foster's iconic Willis Faber Dumas office building in Ipswich, EMI HQ in Tottenham Court Road, London, modernization of the Royal Liver Building in Liverpool and the Royal Liverpool Teaching Hospital (completed under the Bovis System).

In 1980, Bovis was selected as management contractor for Richard Roger's landmark Lloyd's building in London, and this innovative project with the building services and glass lifts on the outside was completed in 1986. In 1981 management contracting represented around half the work load of Bovis, while in 1984 the majority of Bovis's contracts were carried out under this system (Cooper, 2000). In contrast in 2005 over 80% of Bovis Lend Lease UK's workload was executed under a negotiated two-stage lump-sum design and build approach mostly under long-term partnering relationships.

Management contracting requires the contractor to be involved in two stages. During the design phase, which will extend into construction, the contractor's role is to advise the employer on the buildability of the design, plan the construction and discuss cost estimates with the quantity surveyor (QS). During the construction phase, the contractor is responsible for tendering parcels of work and negotiating subcontracts with subcontractors (known as works contractors) on behalf of the employer.

The management contractor then enters into lump-sum contracts with the works contractors after approval by the client. The management contractor is solely responsible for managing the design and construction and supervises the work on site. For this they are paid a fee, sometimes with a bonus if the project is finished to time and budget. In addition, they are paid for site management, site facilities and administration, either on cost-reimbursement basis or by lump sum. Although management contracting sets out to encourage less adversarial attitudes between the various participants, the objective has frequently been nullified by the tendency of the client to assign more and more risk to the managing contractor, risk which they can neither manage nor reasonably price (Centre for Strategic Studies in Construction, 1991). In the 1980s management contracting was used on several major projects, for example, Heathrow Terminal 4 and Manchester Airport Terminal Buildings, Birmingham Convention Centre and the Headquarters for the HSBC bank in Hong Kong etc.

Management contracting is closely identified with the 1980s construction boom. However, it still encountered, albeit in forms heavily amended from the first recognized standard form, the JCT87 Management Contract. However, it is noted that the new JCT 05 suite of contracts contains a Management Build family of contracts.

The services offered by the management contractor may include those given under the sub-headings below:

Panel 11.1 Case study: management contracting – HSBC HQ Hong Kong

The HK$5,000 million (£375 million) Headquarters of the HSBC bank in Hong Kong was designed by Norman Foster/Ove Arup and constructed using the management-contracting approach with John Lok/Wimpey joint venture as construction manager.

This 55-storey iconic tower block in the Central District of Hong Kong, the structure of which resembles a North Sea oil-drilling rig, was completed within a project cycle of six years (design brief 1979, two years pre-contract, four-years construction with a completion in 1985).

Among the reasons for the client selecting the management-contracting approach were the following:

- The need for an early start and early completion of the project in a situation where the design was not sufficiently defined prior to construction. The circumstance requires good planning and control of the design/construction overlap and careful packaging of contracts;
- The need to consider particular construction methods during the design phase;
- In this complex project involving high technology, management contracting would provide greater flexibility for design change than conventional contracts;
- The project was organizationally complex. Typically this may arise from the need to manage and co-ordinate a considerable number of contractors and contractual interfaces and several design organizations;
- The client and his advisers had insufficient in-house building management resources for the project.

Source: Archer, 1985.

Pre-contract

Programming the design, design input on buildability, budget and cost forecasts, advice on financing, cost control, materials procurement and expediting, preparation of tender documents, evaluation of tenders, selection of construction contractors, insurances and bonds – policy and implementation, construction planning and programming, methods of working.

Post-contract

Supervision and control of construction contractors, provision of central services and construction equipment, design of temporary works, quality control, industrial relations – policy and monitoring, costing of variations, and certification of interim and final payments to construction contractors, assessment and monitoring of claims.

The strengths of management contracting can be summarized as follows:

- Allows an early start and completion;
- Time is saved by a more extensive overlap of design and construction utilizing the management contractor's expertise in construction planning;
- It allows flexibility particularly where the programme and design are ill-defined and subject to change;
- Cost savings is achieved through better control of design changes, improved buildability, improved planning of design and construction into packages for phased tendering, keener prices due to increased competition on each package;
- It reduces delays and knock-on effect of claims;
- It is easier to control the selection of construction contractors to those of known ability;
- It avoids adversarial attitudes – leads to a more harmonious relationship between the parties.

Panel 11.2 Case study: management contracting – Stoke-on-Trent Cultural Quarter

The NAO Report 'Progress on 15 major capital projects funded by the Arts Council England' found that of the 13 projects then completed, 4 were finished 12 or more months late; 13 of the 15 projects were over budget with overruns ranging from 1.7% to 58%.

The Cultural Quarter project was an innovative initiative which aimed to regenerate the City Centre in Stoke-on-Trent through the provision of two high-quality arts and entertainment venues.

The two venues to date have been an undoubted success and have stimulated further development. However, the project cost the Council £15 million more than the original budget of £22 million – yielding a total spend of £37 million. The additional costs related to an overspend on the main contract (including professional fees) of £5.6 million; unbudgeted items of £1.7 million and a further £7.8 million relating to an unsuccessful arbitration involving the Council and the project's architects.

The Audit Commission's Report identified that

- the Council did not adequately consider the risks associated with the management contract;
- the appointment of each of the major contractors was characterized by poor practice;
- the Council failed to establish effective arrangements for managing the project;
- reporting to Council members was too shallow to allow a proper consideration of risk.

The Report also noted that

- the Council's top priority was cost certainty – it was working to a fixed budget;

- the management form was considered the only option that would allow a contract to be procured by the deadline to secure £1 million in ERDF grant;
- The poor working relationship on the contract (between the architect Levitt Bernstein Associated Ltd and the management contractor Norwest Holst Construction Ltd).

The NAO Report identifies that the Arts Council has now adopted a new approach – partnership contracting – for a number of building contracts. This approach brings together the design team and the building contractor at an early stage of the project with the aim of reducing costs on the project and securing shared commitment to the project.

Sources: NAO Report 'Progress on 15 major capital projects funded by the Arts Council England' by the Comptroller and Auditor General HC 622 Session 2002–2003: 2 May 2003; Audit commission: District Auditor's Report 'Cultural Quarter Stoke-on-Trent City Council', 22 January 2004.

The weaknesses of the system can be summarized as follows:

- A client may be exposed to a greater risk due to: reliance on a contract cost plan prepared on the basis of incomplete information, late information, failure of works contractors to adhere to time or quality standards (there is no direct contractual link between the client and the construction contractors), delays and subsequent time and cost overruns;
- There is evidence that the overall cost may be greater under this fast-track approach; however this is normally offset by an early completion and the additional letting income or revenue accrued;
- There is a tendency for duplication of administrative and supervisory staff;
- Roles and responsibilities of designer and management contractor for quality control are unclear;
- In practice the management contractor's ability to ensure compatibility between design and construction methods may be limited;
- The potential for grey areas between works contractors is high;
- If the JCT management contract is used in an un-amended format the client is responsible for the knock-on effects of the works contractors, not the management contractor;
- Design may suffer as architect is under time pressure.

11.3   Construction management

The move towards construction management was led in the 1980s by the larger, more experienced international clients and developers who have skills in project management that they can apply directly to each project (see Fig. 11.2).

The projects were often massive (e.g. the £800 million Broadgate complex in the City of London and the £3 billion Canary Wharf development in London Docklands), highly speculative and complex. These developers appointed a strong project management team using American organizational and production methods – in contrast to the traditional UK practices, encouraged innovation, for example, fast-track techniques utilizing off-site fabrication and demanded success! It is claimed that the construction management approach is an ideal process for managing risk, the concept being to identify the risks at the earliest possible opportunity and then proactively manage them.

Employer		
Pre-contract design Conceptual designers architect/engineer/ specialist contractors	*Post-contract design* Architect/engineer/ trade contractors	
Construction manager	*Construction* Construction manager/ trade contractors	
		Contracts JCT2005 (CM) NEC3

11.2 Construction management contracting system.

In recent years in the UK we have witnessed the use of the construction management approach on some high-profile public and private projects including: the £234 million Portcullis House – the accommodation block for MPs at Westminster (1989–2000) designed by Michael Hopkins & Partners and built by Laing Management Ltd; the £431 million Scottish Parliament building in Edinburgh (1998–2004) inspired by Enric Miralles and built by Bovis Lend Lease and the iconic £130 million 'Gherkin' HQ for the reinsurance company Swiss Re in the City of London (1997–2000) designed by Foster and Partners and built by Skanska.

The construction management approach demands a client with commitment and expertise to become involved in the development process, as it is they who would usually appoint the design team, the construction manager and the trade contractors direct (see Fig. 11.2). Under this approach the construction manager acts as the client's agent with responsibility for co-ordinating and controlling all aspects of the project. Construction management is very much a team approach with the client, designers and construction contractors, with the construction manager acting as team leader.

It is recommended that the construction manager is appointed by the client at the same time as the architect/engineer, if not before. The consultant construction manager would normally be required to proactively manage the design team's production of information and the various interfaces between the trade contractors. They would also provide expert advice at both the design and construction stages on construction planning, costs, construction techniques and buildability.

The construction manager has no contractual links with the design team or the trade contractors and provides professional construction expertise without assuming the financial risk; they are only liable for negligence, by failing to perform their role with reasonable skill and care, unless some greater liability is incorporated in the contract.

The construction management system

Strengths

- The construction manager's objectives should be closely aligned to those of the project sponsor – being motivated by the level of the fee and enhancing reputation, not by increasing their own profit.

Panel 11.3 Case study: construction management – Great Eastern Hotel, London

The case of Great Eastern Hotel Company Ltd (GEH) v. *John Laing Construction (JLC)* was heard in the Technology and Construction Court in 2005. The case arose after a luxury hotel in London overran by a year and cost £61 million rather than the £38 million budget. JLC were appointed as construction manager under an agreement which provided that JLC should exercise reasonable skill, care and diligence expected of a properly qualified and competent construction manager and they should ensure that each contractor complied with the obligations under their respective contracts. GEH claimed that JLC had caused the overrun by their mismanagement.

Judge Wilcox held that the construction management agreement required the construction manager to manage the construction of the project but not to accept the principal risks of time and cost which remained with the employer. However, while JLC was not the guarantor of the project, it did owe clear professional obligations to the GEH. JLC had an obligation to manage the project so that risks in relation to time and money were minimized. It was held that JLC had breached this obligation in that it failed to manage the contractors with the degree of care expected from a professional in JLC's position. The court held that GEH was entitled to recover £8.9 million in damages.

Source: Glover, G., *Building,* 11 March 2005.

- As with the management-contracting approach construction management offers a better chance of success with greater flexibility within a tight timescale on a complex project.
- Changes in design can be accommodated later than in some other strategies.
- Construction management allows the client a full and continuous involvement in the project and a greater degree of control – indeed the client is the ultimate decision maker between the designer and manager in order to provide the balance between architecture, technology, time and cost.
- The client has greater flexibility in the appointment of the works contractors, direct payment from the client can result in lower bids and in theory there is a better long-standing relationship between the parties.
- Early completion is possible because of overlapping design and construction (known as fast-track construction).
- Construction management utilizes a team concept and provides early input on buildability, potential site problems, planning methods and costs.
- Construction management promotes non-adversarial relationships and team building.

Weaknesses

- As the client contracts direct with the construction contractors the total risk in the event of failure or dispute lies with the client, there being no intermediary main contractor. These risks can be substantial and include not only the project's costs for delay and disruption but also the trade contractors' delay and disruption of each other.

- Clients will have to contribute a great deal of expertise when undertaking construction management; this may prove to be a daunting experience for inexperienced users of the construction industry.
- The client does not know the overall price at commencement of the works (often based on cost-plan estimate). Indeed any degree of price certainty may not be achieved until all the construction work packages have been let.
- The construction manager assumes no risk other than for negligence.

Panel 11.4 Case study: construction management – Portcullis House, London

Portcullis House provides high-quality accommodation for 210 Members of Parliament and 400 staff together with committee rooms. This challenging project was completed in August 2000 after a construction period of 30 months.

The project was initially delayed for almost a year due to the reconstruction of the Westminster underground station which lies directly beneath Portcullis House and suffered a small further delay of six weeks. The project out-turn cost was £234 million, some 18% greater than the original 1992 forecast but 4% lower than the 1998 budget approved by Parliament.

All main construction contracts were let after competitive tendering. However the client incurred legal and other costs totalling some £10 million after it was successfully sued by an unsuccessful tenderer (US specialist Harmon) for unfair treatment and contravention of procurement regulations in relation to the contract for the fenestration (prefabricated wall and window units).

The recommendations of the NAO Report on the project make interesting reading particularly in light of the Scottish Parliament building which was completed four years later.

1. Recognize the importance of managing the risks associated with innovative design;
2. Establish at an early stage a board of senior officials, chaired for larger projects at the highest level, to oversee the project;
3. Provide appropriate training, advice and support for senior decision makers;
4. Carry out investment appraisals or life-time costing exercises prior to approval;
5. Use value engineering to explore the scope to meet the requirement at lower cost;
6. When monitoring and reporting the likely out-turn cost of projects against forecasts, maintain a clear distinction between forecasts made at the time of the initial decision to undertake the project, and later forecasts incorporating agreed changes in costs;
7. Consider regularly publishing information on the cost of major projects;
8. Ensure that liquidated damages clauses are based on sound estimates of likely costs;
9. Ensure that there is adequate control of professional fees and expenses when selecting and appointing professional advisors;
10. Undertake a review of the building in use.

Source: NAO Report *Construction of Portcullis House, the New Parliamentary Building* (2002).

For further information on the construction management procurement route including its successful use on the £100 million refurbishment of Peter Jones department store in Chelsea, London, see Simon Rawlinson's excellent review in *Building* magazine issue of 1 September 2006.

11.4   Management contracting or construction management?

The CIRIA *Special Publication 81* (Curtis *et al.*, 1991) identifies the key factors to be considered by the client in choosing between construction management and management contracting:

1.   Does the client have the desire and ability to exercise direct contractual control over works contractors?
2.   In which system is the management organization better able to use its skills and experience for the benefit of the client?
3.   Are the works contractors likely to respond better under one system than they would under the other?

The Construction Management Forum Report confirms that the construction management approach is the preferred method for those clients who have the capability and confidence to follow the management path of procurement (Centre for Strategic Studies in Construction, 1991). The Report offers practical guidance in implementing the construction management methodology.

11.5   Reflections on the Scottish Parliament building

'The selection of Construction Management was the single factor to which most of the misfortunes that have befallen the project can be attributed', was the quote which hit the headlines after the publication of Lord Fraser's *The Holyrood Inquiry* into the new Scottish Parliament building.

Lord Fraser took evidence from Colin Carter of Gardner Theobold who set out seven *must haves* for the Construction Management procurement route to work effectively.

1.   An experienced and informed client with an understanding of construction and construction processes;
2.   An experienced and sufficient team with good leadership not forced down the route of just trying to keep the project going and managing change;
3.   Well-defined roles and responsibilities from the start;
4.   An architect who can envisage the whole and the detail at the same time, if retrospective change is to be avoided without resultant ripple effect on trade packages;
5.   Sufficient time up-front in planning, to foster a no-surprises culture and to avoid crisis management;
6.   A very good construction instruction, approval and change process;
7.   An effective and well-managed risk-management process.

Building magazine contains an interesting series of articles debating the merits or otherwise of using construction management on public projects. Ashley Piggott identifies the failed projects in the Arts sector including Sadler's Wells Theatre where the costs went out of control and argues the case for a variant of the design and build approach. He considers that construction management should not be used on the £100 million rebuilding of the Shakespeare Theatre in Stratford-upon-Avon and dissects each of Colin Carter's seven *must haves* arguing that 'nobody is ever accountable under the Construction management route and that cost escalation should come as no surprise' (Piggott, 2004).

Panel 11.5 Case study: management contracting became design and build – Welsh Assembly building, Cardiff

The decision to locate the National Assembly for Wales on the present site in Cardiff Bay was taken in April 1998. An international design competition was held with a brief setting out of a functional specification for the building and expressing a clear desire for an open and democratic building. The competition jury considered designs from six architects and recommended a concept design from the Richard Rogers Partnership (RRP).

The specifics of the brief included the stipulation that the building be exemplar for access, that sustainable strategies and renewable energy systems be implemented throughout, that the building have a minimum 100-years life span, and that, wherever possible, Welsh materials be used throughout.

Other elements included a 610 m^2 debating chamber for 60 to 80 members, 3 committee rooms, offices, a media-briefing room, tea room, members lounge, public galleries and a main hall to act as reception and exhibition space.

The initial price limit set for the new Assembly debating chamber was £12 million including fees and completion was scheduled for early 2002. However, by January 2000 the cost estimate had increased to £22.8 million.

The Welsh Assembly considered that the design and build route was inappropriate because of the novelty and complexity of the design and the need to retain client control. In December 2000 the Assembly appointed Skanska Ltd as the management contractor with responsibility for managing the construction of the building; they would also assist the architect with buildability and constructional issues, and let the works to subcontractors.

Concerns arose over cost increases leading to fundamental disagreements between the assembly and the RRP. In July 2001 the assembly terminated the RRP employment to design the building because of a loss of confidence in the firm's ability to deliver the project within budget and the project was suspended. Francis Graves, construction project managers and cost consultants, were appointed to review the whole building and propose a way forward.

In November 2001 the assembly decided to change the management arrangements for the project. In May 2002 Schal International Management Ltd were appointed as project managers. In July 2003, Taylor Woodrow Construction Ltd were appointed design and build contractor with the remit to develop the existing design and complete the construction work for a fixed price of £41 million excluding VAT. Work recommenced in August 2003 with completion due in September 2005 with a final cost of £60 million.

Note: In February 2002 the adjudicator considered that the RRP were entitled to payment of invoices and interest totalling £448,000 (not the full amount of £529,000 which had been claimed). The total legal fees and expert costs incurred by the assembly in this matter was £267,000.

The Welsh Assembly has escaped the furious rows surrounding the runaway costs of the Scottish Parliament. Even so it needs to be remembered that Rogers was taken off the job for two years before winning reinstatement. 'We were pushed out as we worked through the critical equation of time, cost and quality,' says Harcourt (Rogers' Director in Charge of the project) 'the brief kept expanding, the cost implications were serious and rebounded on us.'

Sources: Wakefield *et al.*, 2004; Bourn, 2002; Binney, 2006.

Colin Carter, partner at Gardner Theobold vigorously defends his corner stating:

> In my view Fraser did not condemn CM per se – among his many recommendations he identified two key elements. The first was the flawed decision to choose CM, as he believed the decision was not fully considered, especially the risk element. Second, he does recommend that in future the public sector does not use CM, but he does not rule out CM in other sectors.

(Carter, 2004)

Lawyer Ann Minogue concludes the debate on the use of construction management at the Scottish Parliament building stating:

> Assuming speed remained a priority; most of us would conclude that construction management was the proper choice of procurement route and no other route would predictably have offered a better solution. But the client should be alive to the risks and manage the project accordingly.

(Minogue, 2004)

11.6   Design and manage

This approach combines some of the characteristics of design and build and the management approach. It allows the client more opportunity to be involved in the design process and to make changes to the design and specification. The approach may be desirable on follow-on contracts, on fast-track projects or where the contractor has some specialized expertise. The contractor is often appointed early, often in competition based on a quoted fee with a build-up of the required preliminaries. All construction work is undertaken by specialist construction contractors.

Two variants of design and management are noted: contractor – in which the project design and management firm takes on trade contractors and consultant – in which the project

---

Panel 11.6 Case study: property developer MEPC

Property developer and manager MEPC's assets comprise eight sustainable business communities around the UK – six business parks outside large towns and two strategically located in city centres with a total value exceeding £1.1 billion.

MEPC uses the procurement route best suited to the needs of a project. For schemes that are well-defined, it uses negotiated, lump-sum, design and build contracts. This route is taken for the vast majority of its work.

For projects that are more difficult to define, and where there is a need to react to change or the tenant has particular needs, the company will use construction management contracts. This procurement route is also applicable for buildings requiring complex laboratories and research facilities.

Source: '50 top clients a building directory', *Building* magazine, Supplement February 2003 www.mepc.com/AboutUs/Introduction.aspx – (accessed 20 March 2007).

management and design firm acts as the client's agent with the trade contractors direct to the client.

11.7   EC procurement rules

The Public Contracts Regulations 2006 and The Utilities Contracts Regulations came into force in England, Wales and Northern Ireland on 31 January 2006. The regulations set out the procedures to be followed at each stage of the procurement process and are intended to ensure that public bodies award contracts in an efficient and non-discriminatory manner and with a view to securing value for money and transparency.

All building and civil engineering works executed by public authorities above €5,278,000 must be the subject of a call for competition by publishing a contract notice in the *Official Journal* of the EC. The regulations seek to modernize the rules, provide for new procurement arrangements such as framework agreements and e-auctions and to take into account social and environmental considerations.

The regulations provide for four procurement procedures:

1.   The open procedure – all interested parties may tender;
2.   The restricted procedure – contracting authority can select whom it will invite to submit tenders from those giving initial responses;
3.   The negotiated procedure – contracting authority can negotiate with one party after giving notice, in exceptional circumstances, the negotiated procedure may be used without any prior publication of the contract notice;
4.   The competitive dialogue procedure – the notice in this case invites requests to participate – a minimum of three participants is required.

The first three have been in for some time; the fourth is introduced by the new regulations. The UK Government currently prefers the use of the negotiated contracts for PFI contracts. The new competitive dialogue may be suitable for complex private finance initiative (PFI) and Public–Private Partnership (PPP) projects but the position remains far from certain.

A contracting authority must award a public contract on the basis of the offer which is either the most economically advantageous from the point of view of the contracting authority or offers the lowest price. The criteria to be used to determine that an offer is the most economically advantageous include the following: quality, price, technical merit, aesthetic and functional characteristics, running costs, cost-effectiveness, after-sales service, technical assistance, delivery date, delivery period and period of completion. Where the contracting authority intends to award a public contract on the basis of the offer which is most economically advantageous, it must state the weighting given to each criterion in the notice or contract documents (Oakes, 2006).

For an excellent review of procurement in the public sector see Simon Rawlinson's article in *Building* magazine, 24 November 2006.

11.8   Achieving Excellence in Construction

In 2003 the Office of Government Commerce issued the *Achieving Excellence in Construction* series of Procurement Guides. This series of guides replaced the *Construction Procurement Guidance Notes* series and reflects developments in construction procurement over recent years. The new series is endorsed by the NAO which recognizes that proactive client leadership and robust project management are successful requisites to successful delivery of construction projects and that procurement of construction should be on the basis of 'whole-life value for money'.

Procurement Guide 06 *Procurement and Contract Strategies* identifies the need to demonstrate a significant improvement in performance against quality, cost and time targets. In order to

---

Panel 11.7 Case study: prime contracting – building down barriers

The 'Building Down Barriers' project, set up in 1997 by the Defence Estates and the Ministry of Defence (MOD), was an initiative aimed at establishing the working principles of supply chain integration into construction.

The idea of supply-chain management (SCM) in construction is to harness the full potential in the entire 'Chain' of suppliers of construction services, products and materials, to deliver best value to the client. The value is measured in terms of performance and cost of the facility over the whole of its life, and the aim is to achieve this while at least maintaining – or enhancing – the profit margins of all involved.

Two projects were selected, at Aldershot and Wattisham, with prime contractors AMEC and Laing, and involved the provision of indoor sports and swimming pools for army garrisons. A research and development group, led by The Tavistock Institute and the Warwick Manufacturing Group, developed and evaluated the supply chain process with its supporting tools and techniques.

The whole concept was based on the concept of setting up long-term relationships aimed at improving the value, improving the quality and reducing the underlying costs; it was based on trust, openness and teamwork.

The project started from the client's statement of need in output terms. The design was then developed by the design team involving the key design constructors and component suppliers with the work managed in clusters. Design solutions were assessed using the through-life cost approach, rather than capital price alone, and was expressed as the net present value and also as a target through-life cost profile over time.

The approach incorporated risk management used in conjunction with value management and whole-life costing principles. Payment was made based on the target-cost approach.

The 'Building Down Barriers' project was highly successful in delivering a range of benefits to customers, including the following:

- Enhanced functionality for clients and facility users;
- Savings in whole-life costs;
- Delivery ahead of programme;
- Predictability of cash flow during design and construction;
- Improved collaborative relationships within the project team.

In addition, benefits to the supply chain included more efficient working processes, a positive atmosphere of team collaboration and greater confidence in design information. It was also highly influential in changing MOD's approach to construction procurement to place a strong emphasis on supply chain integration and single-point responsibility. The work had been taken forward via a series of workshops and training programmes, and is a good example of a joined-up approach to generating change in line with the principles of rethinking construction.

Source: Holti *et al.*, 2000.

achieve these, it is essential that all procuring bodies move towards proper integration of design, construction and operating functions. This will require a move to integrated teams, early supply-team involvement, incentivized payment mechanisms, continuous improvement processes and joint commitment to achieving best whole-life value. Three preferred procurement routes are recommended to which framework arrangements may also add value.

Public–Private Partnerships (PPP) (particularly Private Finance Initiative (PFI))

This approach is only recommended for projects whose capital cost is likely to exceed £20 million, and is created for the provision of services and not for the exclusive provision of capital assets such as buildings.

Design and construct (and where appropriate maintain and operate)

In a 'design and build' contract, the integrated project team is responsible for the design and construction of the facility. The supply team is likely to deliver the greatest performance benefit through innovation, standardization and integrated supply chains, where appropriate output specifications are used.

There may be some circumstances where the design and build procurement route should be extended to cover maintenance and also possibly operation of the facility for a substantial period.

Prime contracting (requires there to be a single point of responsibility – the 'prime contractor' – between the client and the supply chain)

The prime contracting procurement route is mainly practised in the UK where the major supporter is the Ministry of Defence. It generally features the equal collaboration of all stake-holders, to the extent that all contractors, consultants and client representatives work together in partnership to ensure that a cost-efficient and suitable project design solution is achieved.

Prime contracting requires that there be a single point of responsibility (e.g. the prime contractor) between the client and the supply team. The prime contractor needs to be an organization with the ability to bring together all of the parties (the supply team) necessary to meet the client's requirements effectively.

It is common that long-term alliances are formed between all contracting parties and down the supply chain. These alliances are designed to allow a longer-term view to be taken for the different projects undertaken. This is particularly important since many of the projects under this arrangement also involve maintenance obligations which often last for 5 to 7 years after the project has been completed.

Prime contracting usually includes such features as pain/gain share (where the prime contractor as well as the client gains financially by reducing the project costs), target-cost pricing (where prices are agreed on the basis of a reasonable profit for the supply team and value for money to the client) and open-book accounting (where costs are made transparent to the client).

11.9   The NHS Procure 21 guidelines

The NHS spends in excess of £1.4 billion a year on capital investment and has the largest capital procurement programme in the UK Government. In 2002 the NHS introduced Procure 21 – an innovative approach to procurement of construction projects in excess of £1 million embracing the principles of *Rethinking Construction*. After the success of the scheme on two pilot projects twelve Principal Supply Chain Partners (PSCPs) were selected.

The key principles of Procure 21 (NHS Procure 21, 2002) are as under:

- scheme management by an integrated supply chain;
- cost certainty at an early stage;
- time certainty through the use of proactive project management techniques;
- assured construction quality;
- financial security;
- skills development for NHS Trust staff;
- the use of lean construction and collaboration techniques;
- cost improvement through continuous improvement;
- open-book accounting of all transactions;
- increased VAT reclaim possibilities and 'Best Practice' captured across all developments and fed into new projects;
- no need for Trusts to go through lengthy OJEC procurement cycle; the use of NEC ECC target-cost contract with activity schedule.

In 2005, with 38 projects completed under the national framework, The National Audit Office reported that prior to Procure 21 only 28% of projects were completed on time, and 26% had a cost overrun. In comparison under Procure 21 all projects had been completed within the agreed budget. The NAO further reported that VFM gains of around 10% against the costs of projects have been achieved using Procure 21 compared to the costs of traditional projects (NAO, 2005).

In summary the advantages of the NHS Procure 21 can be summarized as follows:

- predictable;
- develops value for money;
- encourages innovation/value management;
- enables prompt agreement of costs and start on site;
- single-cost advisor is used.

Possible disadvantages may include the following:

- Procure 21 not mandatory so Trusts may choose traditional approach, in effect this means that the PSCPs may not secure expected volume of work;
- open and collaborative approach is not consistent all the way down the supply chain;
- finance cannot be raised under this approach (unlike PFI);
- higher consultant fees may be incurred;
- skills of successful project directors may be lost if they are retained by the same Trust upon completion of major works.

11.10   Highways Agency – overlying principles for future procurement

In addition to the need to comply with legislation and government policy there are a number of basic principles which can be applied to all categories of work to achieve best value. To be fully effective all of the principles need to be applied as a package when procuring a product or a service. Each of the principles could be adopted in isolation but in total they provide suppliers the structure within which to identify optimal solutions and the incentives to deliver continuous improvement over long-term periods. The principles are set out below:

- early creation of the delivery team;
- an integrated and incentivized supply chain;

- maintaining a competitive and sustainable supply chain;
- clear points of responsibility with no unnecessary layers of supervision;
- E-procurement;
- selection of suppliers on the basis of best value, that is, the optimal combination of quality and price;
- fair allocation of risks;
- high-quality design;
- partnership approach based on long-term relationships;
- performance measurement with continual improvement targets.
  (www.highways.gov.uk/business/932.aspxV [accessed 20 September 2006])

11.11   The 2012 London Olympics

The 2012 London Olympics is one of the biggest challenges facing the UK construction industry for a generation; in terms of scale it is at least twice the size of BAA's Heathrow Terminal 5. The project is located in the Lower Lea Valley in the East of London and has required a massive clean-up of 200 hectares of contaminated land, the removal of pylons and the relocation of power lines underground.

In September 2006 the Olympic Development Authority (ODA) appointed CLM – a consortium comprising Laing O'Rourke, Mace and US consultant CH2M HILL – as delivery partner with the primary role to oversee the rest of the supply chain. Significantly both Laing O'Rourke and Mace are involved in the construction management of the flagship Terminal 5 project.

CLM will also provide resource management, technical capability and systems to manage the planning, design, procurement and delivery of the construction for the venues and infrastructure. It will manage risk and opportunity but the ODA will always remain the contracting authority (*Building*, 2006).

After the appointment of CLM the next stage of procurement was the design and build of the main stadium on which a consortium including Sir Robert McAlpine and HOK has been appointed as the preferred bidder.

Panel 11.8 2012 London Olympics construction commitments

On 3 July 2006 the Strategic Forum for Construction introduced the '2012 Construction Commitments'. The aim of the document, developed by industry with the strong support of government, is to maximize the opportunity to showcase the very best of British construction practices, using the Olympics as a live example.

The '2012 Construction Commitments' covers six key areas of the construction process and is designed to promote collaborative working and best practice, ensuring the successful delivery of the Games infrastructure, buildings and subsequent legacy. The document does not involve any new initiatives but strives to make the most of existing initiatives, tools and talent in the industry.

The Commitments was developed by the Strategic Forum for Construction's 2012 Task Group in conjunction with the Department of Culture, Media and Sport and the Department of Trade and Industry and cover six areas including

- client leadership;
- procurement and integration;
- design;

- sustainability;
- commitment to people;
- health and safety.

The commitment on procurement and integration

A successful procurement policy requires ethical sourcing, enables best value to be achieved and encourages the early involvement of the supply chain. An integrated project team works together to achieve the best-possible solution in terms of design, buildability, environmental performance and sustainable development.

- Procurement decisions will be transparent, made on best value rather than lowest cost, use evaluation criteria and where appropriate, specialist advisors, while encouraging the contribution of smaller organizations;
- All members of the construction team will be identified and involved at an early stage, particularly during the design process, and encouraged to work collaboratively;
- Supply chain partners will be required to demonstrate their competency, their commitment to integrated working, innovation, sustainability and to a culture of trust and transparency;
- To ensure effective and equitable cash flow for all those involved, all contracts will incorporate fair payment practices, such as payment periods of 30 days, no unfair withholding of retentions, project bank accounts, where practicable and cost-effective, and will include mechanisms to encourage defects-free construction;
- The duties of each project team member will be identified and shared at the outset of the project and appropriate insurance policies, such as project insurance, put in place;
- Risks will be clearly identified, financially quantified and allocated in line with each party's ownership and ability to manage the risk;
- All contracts will have an informal and non-confrontational mechanism to manage out disputes;
- The employment practices of all organizations, including subcontractors and the self-employed, will be scrutinized by the client and the supply chain to avoid abuses.

Source: www.strategicforum.org.uk/2012CC.shtml (accessed 6 December 2006).

11.12   Selecting the procurement route

Selecting the most appropriate procurement route is one of the most crucial decisions taken on any construction project. All routes have their advantages and disadvantages. What is needed is an objective appraisal of the alternatives.

Achieving Excellence in Construction recommends that the following questions are asked about the contract strategy:

- What resources and expertise does the client have?
- What influence/control does the client need to exert over design?
- Who is best able to carry out the design?

- What influence/controls does the client wish to exert over the management of the following:

  a) Planning (project, construction)?
  b) Interfaces (project, end-users)?
  c) Risk?
  d) Design?
  e) Construction?

- What can the market provide and what framework agreements are already in place?

The *Achieving Excellence Procurement Guide 06 Procurement and contract Strategies*, OGC, 2003 also includes the following checklist:

Assessing the procurement route (see Table 11.1):

- Is this the right procurement route for the project, backed up with a contract in which the roles and responsibilities are clearly defined?
- Are choices about allocating risk and control tailored to the circumstances of the project and reflected in the procurement strategy?
- Has the most appropriate integrated procurement route been chosen – PFI, design and build or prime contracting?

Assessing the contract:

- Have improvement targets and measurement arrangements been agreed with the integrated project team and quantified?
- Have incentives been included in the contract to encourage the integrated project team to perform well and achieve the client's objective?
- Have the required benefits been quantified before incentive payments are paid?

11.13   Achieving Excellence in Construction methodology

Table 11.1 (at the end of this section) shows the methodology recommended by the OGC's *Achieving Excellence in Construction*. The challenge is to select a procurement route which delivers best value for money to the client. The evaluation criteria is selected which is appropriate to the client and the project; such issues as supplier innovation, whole-life costing and sustainability should be considered.

Each criteria is then given a weighting out of a total of 100. Each procurement route is then assessed in turn by the client's project management team awarding a score against each criteria. The individual scores are then multiplied by the criteria weightings to arrive at the final scores for each procurement route.

11.14   Conclusion

This chapter has reviewed the two main management methods of procurement – management contracting and construction management. The strengths and weaknesses of each have been clearly identified. The successes of the CM approach in the private sector, for example, the Honda car plant at Swindon and the Peter Jones retail store in London clearly demonstrate that this strategy has a future in the hands of expert clients. By contrast, the Scottish Parliament building and the Stoke-on-Trent Cultural Quarter projects have highlighted the potential problems that might arise if these strategies are used in the public sector.

Table 11.1 Illustrative example of procurement route VFM evaluation mechanism.

Project Title: University of Metropolis student accommodation

Procurement route	Criteria weight %	Traditional		Design and construct		Design build and maintain		Design build, maintain and operate		Prime contracting	
Evaluation criteria (Appropriate to the client and project)		Score	Weighted score	Score	Weighted score	Score	Weighted score	Score	Weighted score	Score	Weighted score
Opportunity for supplier to innovate to yield the most cost-effective combination of capital construction, maintenance and operation	15	5	0.75	50	7.50	70	10.50	90	13.50	80	12.00
Least disruption in project flow due to perceptions and procedures to meet public accountability – minimization of disputes	15	5	0.75	50	7.50	70	10.50	90	13.50	70	10.50
Certainty of whole-life costs	10	5	0.50	50	5.00	70	7.00	90	9.00	80	8.00
Flexibility for future changes in client requirements and post-completion change	10	90	9.00	75	7.50	60	6.00	50	5.00	70	7.00
Speed of project delivery to occupation/first use	5	25	1.25	50	2.50	60	3.00	70	3.50	25	1.25
Control over detailed design and design quality (a detailed output specification is still required)	10	95	9.50	70	7.00	60	6.00	50	5.00	90	9.00
Control over whole-life health and safety issues	5	5	0.25	50	2.50	60	3.00	70	3.50	60	3.00
Reduction in disputes and in-house costs through single-point responsibility	5	5	0.25	50	2.50	70	3.50	90	4.50	80	4.00
Control of sustainability issues Requirement to optimize whole-life cost (incl. above)	25	95	23.75	70	17.50	60	15.00	50	12.50	90	22.50
Total scores	100		46.00		59.50		64.50		70.00		77.25
Order of how well the procurement route fits the evaluation criteria		5		4		3		2		1	
Overall order of priority (i.e. combination of total score and best fit)		5		4		3		2		1	
Members of Evaluation Panel Panel member 1.................Signature Panel member 2.................Signature											

Source: Developed based on *Achieving Excellence in Construction, Procurement Guide 06, Procurement and Contract Strategies*, 2003.

The chapter has also highlighted the significance of the *Achieving Excellence in Construction* Procurement Guides. The chapter also discusses two public sector employers, the NHS and the Highways Agency, who have embraced these recommendations with considerable success.

The chapter finally concludes with a case study based on selecting an appropriate procurement route for a university building using the *Achieving Excellence in Construction* guidelines. In the example (Table 11.1) the recommended strategy is the Prime Contracting route as this scores highest based on the client's chosen evaluation criteria.

11.15   Questions

1. The University of Metropolis is keen to replace its whole portfolio of student accommodation. The new requirement is for a mix of modern accommodation for 10,000 students on three campuses built over a seven-year period. The Pro Vice Chancellor is particularly keen that the project demonstrates value for money (VFM) and embraces the key sustainability issues as follows:

    a.   As the client's project manager recommend a suitable procurement strategy for this project.
         In the solution shown in Table 11.1 *Prime Contracting* option scores 77.25 and is the recommended strategy.

2. A major international car manufacturer wishes to extend its existing car plant with a second 50,000 m² car plant. The project comprises a combination of heavy civil engineering involving deep excavations, steel-sheet piling, bored concrete piling, heavy reinforced concrete foundations, a steel frame with hanging conveyor plant and extensive infrastructure works. The work also comprises a complex array of mechanical, electrical and process services machinery.

    The challenge of the project is the co-ordination of numerous parallel activities on a fast-track programme, while achieving the flexibility of finalizing many automotive processes and robotic installations at the latest possible time.

    The client wishes to be actively involved in the project and to secure best value.

    Recommend an appropriate procurement strategy and identify the key issues required to be considered in order to achieve a successful outcome for the client.

Solution

This question is based on the paper '*Insights from beyond construction: collaboration – the Honda experience*' by Richard Bayfield and Paul Roberts presented to the Society of Construction Law in Oxford on 15 June 2004 (www.scl.org.uk [accessed 2 August 2003]).

Bibliography

Archer, F.H., Project Director John Lok/Wimpey JV (1985) Private correspondence with the author
Bennett, J. (1986) *Construction Management and the Chartered Surveyor*, RICS
Binney, M. (2006) 'A vision of sea, sky and cedar', *The Times*, Monday 13 March
Bourn, J. (2002) 'The National Assembly's New Building: Update Report', prepared for the Audit General for Wales by the National Audit office Wales
Building (2006) 'Delivering 2012', *A Building magazine Supplement*, 17 November 2006
Carter, C. (2004) 'Don't twist my words', *Building* magazine, 22 October
Centre for Strategic Studies in Construction (1991) *Construction Management Forum Report and Guidance*, University of Reading

Construction Industry Research and Information Association (CIRIA) (1983) 'Management Contracting', CIRIA Report 100, CIRIA

Cooper, P. (2000) *Building Relationships: The History of Bovis 1885–2000*, Cassell & Co.

Curtis, B., Ward, S. and Chapman, C. (1991) *Roles, Responsibilities and Risks in Management Contracting, Special Publication 81*, Construction Industry Research and Information Association (CIRIA)

Glover, G. (2005) 'Whatever happened to those fearless construction managers?' *Building* magazine, 11 March

HM Treasury, Central Unit on Purchasing (CUP) (1992) *Guidance Note No. 36 Contract Strategy Selection for Major Projects* (since withdrawn) HM Treasury

HM Treasury, Central Unit on Purchasing (CUP) (1992) *Guidance Note No. 51 Introduction to the EC Procurement Rules*, HM Treasury

HM Treasury, Central Unit on Purchasing (CUP) *Procurement Guidance No. 1 Essential Requirements for Construction Procurement*, HM Treasury

HM Treasury, Central Unit on Purchasing (CUP) *Procurement Guidance No. 2 Value for money in Construction Procurement*, HM Treasury

Holti, R., Nicolini, D. and Smalley, M. (2000) *The Handbook of Supply Chain Management: The Essentials Building Down Barriers, C546*, CIRIA

Minogue, A. (2004) 'Seduced by simplicity', *Building* magazine, 15 October

Nahapiet, J. and Nahapiet, H. (1985) *The Management of Construction Projects: Case Studies from the USA and UK*, CIOB

National Audit Office (2005) *Improving Public Services through Better Construction, Report by the Comptroller and Auditor General*, HMSO

NHS Procure 21 (2002) 'NHS Procure 21', NHS Estates Website www.nhs-procure21.gov.uk (accessed 18 November 2006)

Oakes, R. (2006) 'The public contracts regulations bring in new rules for public procurement', *Building* magazine's Housing and Regeneration Supplement, February

Office of Government Commerce (OGC) (2003) *Achieving Excellence in Construction, Procurement Guide 06 Procurement and Contract Strategies*, OGC

Pigott, A. (2002) 'To D&B or not to D&B?', *Building* magazine, 18 January

Pigott, A. (2004) 'He knew he was right', *Building* magazine, 8 October

Rawlinson, S. (2006) 'Procurement public sector projects', *Building* magazine, 24 November, pp. 52–56

Rawlinson, S. (2006) 'Procurement construction management', *Building* magazine, 1 September, pp. 62–67

Wakefield, S., Oag, D., and Burnside, R. (2004) 'The Holyrood Building Project', SPICe briefing, The Scottish Parliament

Useful website

www.hm-treasury.gov.uk

12 Payment systems and contract administration

12.1 Introduction

Payment systems can be classified in a variety of ways and any classification is unlikely to be exhaustive. Contract strategies can be broadly categorized as either price-based or cost-based.

1. Price-based – lump-sum or re-measurement with prices being submitted by the contractor in their bid, or
2. Cost-based – cost-reimbursable or target-cost, the actual costs incurred by the contractor are reimbursed together with a fee to cover overheads and profit.

A key consideration in the choice of payment system is the allocation of the risk to the parties. Fig.12.1 shows the different payment systems identifying the risks attached thereto.

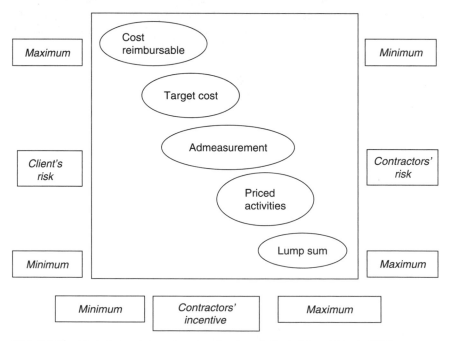

12.1 Relationship between types of payment systems (adapted from Ridout, 1982).

Thomas Telford, one of the greatest engineers in Victorian Britain, designed and supervised the construction of over 900 miles of new roads and bridges in the remote Scottish Highlands in the early 1800s. Papers deposited in the Institution of Civil Engineer's library indicate that this 20-year-long project was split into sections based on lump-sum, fixed-price-based contracts with the contractors taking all the risk. The papers indicate that after encountering considerable unforeseen ground conditions several of the contractors claimed for additional expenses explaining that without this they would be forced into bankruptcy with their workforce and families reduced to starvation. Telford allegedly was unmoved and refused to pay any extra.

Over the past decade in the UK the building sector has been moving away from the use of the lump-sum approach based on Bills of quantities (BofQ) towards the lump-sum plan and specification approach, in which each tenderer evaluates the work using their own builders' quantities. Likewise the civil engineering sector has moved away from admeasurment approach to priced activity schedules linked to the construction programme or the target-cost approach as contained within the NEC *Engineering and Construction Contract*.

The BofQ system had been in use for over one hundred years and if correctly used had many positive features. The BofQ approach demanded that the works were substantially designed prior to tender. However, in practice, the BofQs were often prepared based on incomplete information, a situation that created a false sense of security for the client who carried the risk for any errors or omissions in the original measurement. Furthermore, on these contracts if the contractor was failing to perform, the main sanction available to the client was the incorporation of liquidated damages where the whole or specified sections, of the work are completed late. This sanction is often too late to have an effective influence on the contractor's performance where delays are occurring, as they frequently do, during mobilization and ground works which occur early in the programme for construction.

12.2   Price-based, lump-sum plan and specification

Under the lump-sum system contractors are required to estimate the quantities and subsequently calculate the tender sum based on the client's design drawings and specification. The design should therefore be completed prior to tender with little or no changes to the design anticipated after tender. Lump-sum contracts should thus provide a client with maximum price certainty before construction commences.

The payment to the contractor can either be on fixed instalments or linked to the progress of the works. Schedule of rates may be used under this system to facilitate valuation of variations. The schedule seldom attempts to be comprehensive and if the rates are not made part of the tender can be unreliable.

Seeley (2001) considers that plan and specification contracts are suitable for small works where BofQs are not considered necessary. He highlights the potential danger of using the plan and specification approach: 'On occasions this type of contract has been used on works that were extremely uncertain in extent and in these circumstances this represents an unfair and unsatisfactory contractual arrangement.'

Seeley (2001) further observes that the NJCC have also identified the following difficulties which are likely to arise when using this system for:

1. effectively comparing and evaluating tenders when each contractor prepares their own analysis;
2. accurately evaluating monthly payments;
3. accurately valuing variations;
4. maintaining proper financial management of the contract.

Table 12.1 Trends in main methods of procurement in the building sector – by number of contracts.

		1995	1998	2001	2004
By number of contracts	Lump sum – firm BofQ	39.2	30.8	19.6	31.1
	Lump sum – spec and drawings	43.7	43.9	62.9	42.7
	Lump sum – design and build	11.8	20.7	13.9	13.3
	Target contracts				6.0
	Re-measurement – approx BofQ	2.1	1.9	1.7	2.9

Source: RICS (2006).

Despite the potential disadvantages in using this system it has become the most popular arrangement used in the building sector accounting for 43% in number (see Table 12.1).

12.3   Price-based, bills of quantities (BofQ)

The traditional procurement system in the UK requires the production of a BofQ, which is normally executed by the client's quantity surveyor (QS). These detailed BofQs are prepared in accordance with the rules as stated in the appropriate Method of Measurement and reflect the quantities of designed permanent work left by the contractor on completion. 'The bills of quantities is designed primarily as a tendering document, but it also provides a valuable aid to the pricing of variations and computation of valuations for interim certificates' (Seeley, 1997).

When a BofQ is used it usually forms one of the contract documents and the client carries the risk of errors. In contrast, where the contractor computes the quantities, the contractor takes the risk of errors in the quantities.

Within the building sector the BofQ was prepared in accordance with the highly detailed rules contained within the *Standard Method of Measurement of Building Works*, 7th edition (SMM7). This system is classified as a lump-sum contract and is not subject to re-measurement, it is only subject to adjustment in the case of variations, claims, fluctuations, prime cost and provisional sums etc. In 2004, it was estimated that approximately 24% by value of all building work is executed using BofQ prepared based on the rules contained in the SMM7.

In contrast, in the civil engineering work sector the BofQ is prepared (see Table 12.2) based on one of two methods of measurement – the Method of Measurement for Highway Works (MMHW) for highway works and CESMM3 for other civil engineering works. Due to the unforeseen nature of the ground conditions, the BofQs were based on estimated quantities with the whole work subject to re-measurement based on the finalized drawings or in the case of excavation of rock-agreed site levels. The administration of major projects based on the admeasurement process is extremely staff intensive with teams of QS/measurement engineers (representing both parties) typically, more than three weeks in each month involved in the re-measurement required for the interim valuations. Indeed, the effort in site measurement and valuation diverts the commercial team from what should be considered the most important tasks – those of valuation of variations, early resolution of claims, final measurement and ascertainment or accurate prediction of the total cost of the contract (Gryner, 1995).

The Scottish Office's Roads Directorate were one of the first major clients in the civil engineering sector to react to the growing adversarial and commercial reality of the re-measurement approach when using the ICE 5th edition of *Conditions of Contract*. In the early 1990s the Scottish Directorate introduced an alternative tendering initiative for their roads and bridges projects based on a design/refine and build partnering approach.

Table 12.2 Example of bill of quantities (BofQ) prepared using CESMM3.

Number	Item description	Unit	Quantity	Rate	Amount
	In-situ concrete				
	Provision of concrete				
F143	Standard mix ST4 cement to BS 12,				
	20 mm aggregate to BS 882	m³	840		
	Placing of concrete				
	Reinforced				
F634	Suspended slab thickness:				
	exceeding 500 mm; voided bridge				
	deck	mm	462		
F654	Columns and piers cross-sectional				
	area: 0.25–1 m²	m²	96		
F664	Beams cross-sectional area:				
	exceeding 1 m²	m²	38		

Comments on the perceived merits of the BofQ system

1. *Prompts the client and design team to finalize the design before the bill can be prepared*:

    One of the main advantages of the BofQ system is that in theory it should force the design team to identify ambiguities in the contract documentation prior to tender. However, in practice, the BofQ may be prepared from partially completed design information. In effect, the BofQ may conceal an absence of preplanning and investigation. Indeed, the quantities may be overestimated if the design is incomplete.

    The employer could be exposed to additional risk if there are errors or omissions in the BofQ (e.g. see ICE 7th clause 56(2)). Essentially, the employer guarantees accuracy of quantities in the BofQ. If the drawings are not fully defined then the contract should not be based on a BofQ.

2. *Avoids the need for all contractors to measure the works themselves before bidding and avoids duplication of effort with resultant increase in contractors' overheads which are eventually passed onto clients*:

    BofQs are extremely detailed particularly under SMM7. However, 80% of the cost is usually contained in 20% of the items (Pareto Rule). Despite being detailed few BofQs deal satisfactorily with mechanical and electrical work, which can equate to between 30–50% of the total value of the building. Preparing BofQs also increases the time to prepare the tender documentation.

3. *Provides a commonality in tenders thus providing the opportunity for realistic tender evaluation*:

    The BofQ system allows contractors the opportunity to identify those items, which are under- or over-measured and then load or lighten the appropriate rates. Additionally, contractors may front-end load the rates in the early work sections (in order to improve cash flow). Indeed, the learned QC Ian Duncan Wallace considered that 'the premium for survival as an engineering contractor today must rest as much upon skill in exploiting the profit opportunities afforded by the contract documents (generally known as "loop-hole engineering") as in the efficiency of the control and management of the project itself' (Wallace, 1986). The employer may attempt to reduce high rates in the lowest tenderer's bid prior to finalizing the contract, thus breaching the parity of tender guidelines. In practice the lowest bid may in fact be too high a risk for the employer.

4. *The unique coding system identified in the method of measurement against each item enables contractors to utilize computers efficiently for estimating BofQs*:

Again true in theory, but one contractor may be bidding for work with BofQs based on many different methods of measurement, for example, SMM7, CESMM3, MMHW and International etc. Furthermore, the same BofQ description may be required to be priced differently due to different cost considerations, for example, due to different temporary works.

5. *Can be used as a basis for monthly interim valuations*:

Agreed, but in practice interim valuations may become too detailed and thus require a vast number of man-hours to produce (particularly on re-measurement contracts) Interim valuations can be based on a typical 'S' curve prescribed by the client in the contract documentation – as GC/Works/1 which was based on the analysis of over 500 projects by the Property Services Agency (PSA). However, Klein (1995) noted that the GC/Works/1 system tends to get into difficulties when applied to subcontracts, where the throughput of work and materials can vary from trade to trade.

6. *Rates contained in the bills can be used as a basis for the valuation of variations*:

True they can, indeed both standard forms JCT05 and ICE7th positively require this as a starting point. However, before even the simplest pro-rata rate is calculated the employer's QS needs to have a detailed breakdown of the contractor's unit rates which is not normally available in the UK. It is understood that contractors are required to submit a complete analysis of their tender rates under the German VOB and the Singapore private sector contract.

In practice there are many occasions when the BofQ rates are not appropriate for valuing variations. Variations can involve: delays, interruptions, loss of productivity, out-of-sequence working, uneconomic use of labour and plant, additional supervision etc. The Employer's QS looks to impose BofQ rates, the contractor's QS looks for full recovery based on cost – significant chance of dispute. The best approach: parties agree on the value of variations before the work is executed – including the costs of delay and disruption (this method is now becoming the standard approach following its introduction into the NEC ECC Contract, the JCT05 and the ICE7th).

7. *Can assist the parties in the control and financial management of the works*:

Pre-contract estimating: On building works the client's QS uses data from previous projects to establish the budget price. However, the Building Cost Information Service (BCIS) format requires the BofQ to be in elements. This approach is not favoured by contractors as they require sections in trade format for subcontractors' enquiries.

Post-contract contract control: BofQ describes the permanent works or materials left on completion; they do not reflect the contractor's significant items of cost, for example, labour gangs and construction equipment retained on site for continuous periods (tower cranes, scaffolding). Contractors often have a monthly cost-monitoring system (costs compared to value) – often too late to do anything about identified variances. The BofQ system may in fact divert contractors from developing effective cost-control systems.

There has been much research work on the post-tender use of BofQs. Skinner (1981) considered that although the existing BofQs make a substantial contribution to the post-tender work, they are not ideally suited, either in format or content, to the needs of tendering or production. He considered that the principal reason for the inadequacy of the BofQ, both as a tender document and as an information source to contracting lies in the fact that the requirements of production are not considered. Critically contractors must consider both the time and methods reflecting their use of heavy plant and mechanical equipment. However, Skinner (1981) found that the most widespread criticism concerned the use of 'prime cost sums' to cover nominated subcontractors – the contractor having no involvement in their selection although contractually responsible for them.

Indeed BofQs might lull clients into a false sense of security and provide contractors with an opportunity for increasing profits. However, the BofQ has been used satisfactorily on many major projects and is still used as the basis for calculating the bids by design and build contractors.

12.4   Operational bills

In the 1960s the Building Research Station (BRS) (later known as the Building Research Establishment) identified that the traditional BofQ essentially described the completed works namely, labour and materials. Bills of quantities did not consider some of the main elements of the contractor's costs, for example, construction plant and temporary works, and interim payments made based on work completed were thus arbitrary.

The BRS thus introduced an alternative form, called operational bills, which is much more closely related to the process of construction and would thus provide effective communication between design and production (BRS, 1968).

The development started from the consideration that if the process cost, the part of the tender on which the builder's offer is usually won or lost, is to be realistically priced, labour plant and overhead costs must be estimated on a time basis by reference to an online construction programme (Skoyles, 1968). In this format the BofQ was subdivided into two main sections:

- site operations described in the form of schedules of materials, labour and plant requirements;
- work prefabricated adjacent to or off site.

Although a precedence diagram was provided, it was not an instruction on how the work should be carried out (BRS, 1968). The contractor remained free to adopt any method of construction in relation to his resources (Skoyles and Lear, 1968). It was also noted that the precedence diagram giving the operations could be quickly converted into critical-path analysis by the successful contractor (BRS, 1968).

Operational estimating required the architect to produce operational drawings, pre-judging the contractor's methods and order of working. The operational bills were both bulky and costly to produce and radically changed the estimating process, resulting in increased work for the contractors' estimators.

Shortly afterwards a half-way house version of the operational bill was introduced by the BRS, initially called *Activity Bills* (Skoyles, 1968) but later renamed *Bills of Quantities* (operational format). This approach followed the philosophy of the operational bills except that the operations were detailed in the bill in accordance with the rules of the Standard Method of Measurement rather than labour and materials (Skoyles, 1968).

These were essentially experiments aiming at searching for ways in which the BofQs could become more useful as a key financial document of control. By linking the analysed price to the construction programme these innovative bills highlighted the need to link the two key documents. In practice this approach might have created more problems than it solved particularly in relation to the client's responsibility for the accuracy of the operational bill and network programme!

On reflection this approach was way ahead of its time. However, the idea does seem remarkably similar to the concept of priced-activity schedules (Table 12.4) under the NEC *Engineering and Construction Contract* introduced nearly thirty years later, the key difference being that the contractor prepares this information – not the client's QS.

12.5 Price-based, method-related bills

Method-related bills were first recommended by Barnes and Thompson (1971) in their research *Civil engineering Bills of Quantities*. The authors identified the shortcomings of the traditional BofQ which stem primarily from its failure to represent the effect of the methods of working and timing upon costs and subsequent difficultly in financial control.

When using the Civil Engineering Standard Method of Measurement (now CESMM3) the BofQ included a method-related charges section – in practice several blank pages. This allowed the tenderers the opportunity to insert those major items of construction equipment/temporary works/supervision/accommodation etc., which were not directly quantity related. Method-related charges were identified as either *time-related charges* (e.g. operate tower crane during the project) or *fixed charges* (e.g. bring tower crane to site and set up).

It is claimed, that when the method-related charges approach is used correctly, payment is more realistic and there are fewer disputes regarding changes. However, in practice many contractors chose not to reveal their method-related charges in their tenders, thus cancelling out any advantages of this approach with a preference to continue the commercial game.

12.6 Price-based bills of quantities (BofQ) with milestone payments

In 1994 the Latham report recommended that all contracts have *an* 'express provision for assessing interim payments by methods other than monthly valuation i.e. milestones, activity schedules or payment schedules' (see Table 12.3).

The Hong Kong Mass Transit Railway Corporation (HKMTRC) had long recognized the advantages of linking payments to the achievement of milestones. Under their approach on the HK Island Line (1979–1984) the BofQ was split into major cost centres (between four and six per project) including one covering the mobilization/preliminaries items.

Each contractor was required to submit an *interim payment schedule (IPS)* with his bid detailing the percentage of each cost-centre value they wished to receive during each month of the project. Subject to satisfactory progress, measured by achievement of milestones predetermined by the client, payments were made in accordance with the schedule.

Milestones are clearly defined objectives usually on the critical path for each cost centre and are usually set at every three months for each cost centre. The milestone schedule was included in the particular specification for the project. In the event that a milestone was not achieved payment was suspended on that cost centre until achievement of the milestone is achieved; there was therefore compelling financial reasons for adherence to the programme.

The HKMTRC's bespoke contract on this project followed the general principles contained within the *FIDIC Conditions of Contract for Works of Civil Engineering Construction* (4th Edition, 1987). The modified conditions allowed changes to be made to the cost-centre values during

Table 12.3 Extract from a typical completed interim payment schedule (months 1–5 in a 36-month contract).

Month	Cost centre A	Cost centre B	Cost centre C	Cost centre D
1	1.00	1.00	2.00	0.00
2	2.00	3.00	5.00	0.00
3	3.50	5.00	10.00	0.00
4	5.00	9.00	15.00	1.00
5	7.00	12.50	20.00	2.00

Source: Potts, 1995.

the course of the works for interim payment certification purposes to take account of the following:

- re-measurement of the works;
- valuation of variations;
- claim for higher rate or prices than original rate;
- deduction or reintroduction of the value of goods and materials vested in the employer;
- works instructed against prime cost items and provisional sums.

Russell (1994) reports that on the Hong Kong Airport Railway project the conditions of contract allowed the engineer to make a review of the milestone schedule or IPS if the following conditions were met:

- approval of a revision to the programme which involved a significant change to the sequence and timing of the works;
- award of an extension of time or notification of an earlier date for achievement of a completion obligation or milestone;
- the engineer's instructions for delay-recovery measures in accordance with the provisions of the contract;
- the suspension of works or part of the works;
- significant change to the value of the cost centre.

By using this system the HKMTRC gained the following benefits:

- the ability to compare the true amount of tenders by using the discounted cash-flow method by evaluating tenders on a net present value basis – in effect penalizing contractors who had front-end loaded their tenders;
- identification of the project cash-flow liability;
- senior management's ability to monitor progress simply by monitoring the anticipated and actual cash flows;
- obtaining contracts which positively motivated contractors to meet key milestone dates and complete the work on time – thus providing a link between performance and payment and rewarding the efficient contractor;
- reduction in QS staff needed to prepare monthly valuations;
- greater likelihood of completion within time and cost, thus obtaining best value for money.

Under this type of contract the construction programme is of paramount importance, specifying the level of detail required and the frequency of submissions and reports. The contract also requires the engineer's approval, rather than merely consent, to each programme. Gryner (1995) further identified the increasing trend to employer to specify to the tenderers the type of software to be used (Primavera P3 on the HKMTRC, and the programme logic from which the cost centres and milestones are derived). Any contractor tendering on the basis of different logic could provide alternative milestones and dates for their achievement, but these should be agreed before the award of the contract.

Potts and Toomey (1994) identified that the HKMTRC IPS system worked better on *linear* projects and had limitations on complex building work. They also identified that the success, or otherwise, of alternative payment systems can be directly attributable to the reasonableness and promptness of the client and the client's contract administrator in operating the scheme.

David Sharpe, the former chief civil engineer of the KMTRC considered that 'This payment system and its link with the Milestone Dates provided everyone (and particularly the construction management) with a clear and unequivocal measure for progress, and assisted management efforts towards the aim of completion on time' (Sharpe, 1987).

It is noted that this same IPS payment system was used on the £3.5 billion London Underground Jubilee Line Extension (JLE) which was completed in 1999. Mitchell (2003) identifies that the milestone system worked well for the early stages of the contracts, but the milestones became somewhat *toothless*, as the contractors submitted working programmes substantially different from their tender programmes.

Bob Mitchell further identifies that there were so many changes to programmes caused by delay and disruption, coupled with the award of extensions of time and acceleration measures, that the milestones had to be revised anyway. Despite this, payments still had to be made to contractors regardless of their progress against the milestones and many have commented that the JLE system fell into disrepute.

In the building sector the closest to the milestone approach has been the provisions within the JCT98 *With Contractor's Design*, which provides for payments to be made at the completion of certain stages of the work. However, Klein (1995) identified two possible difficulties in stage-based payments: scope for dispute over whether the work has been completed, and risk that completion of each stage may be delayed by others which could mean no payment at all, particularly for specialist contractors.

12.7   Price-based activity schedules

Historic use

The use of activity schedules for payment purposes is not a new concept. The author recollects that the HKMTRC used this approach on Stage 1 (Modified Initial System, 1975–1980). These major contracts comprising the Nathan Road underground metro stations and tunnels were let on a design and build basis. Contractors were required to submit a lump-sum tender broken down into major activities listed in the tender documentation.

Interim payments were made against completed activities or pro-rata thereto based on a percentage approach calculated based on a simplistic measurement; a brief schedule of rates was also submitted by the contractors with their tenders which were to be used for the valuing of any variations. For interim payments the scheme worked well but was found lacking when valuing variations particularly for items not on the schedule.

The NEC

The Engineering and Construction Contract (NEC3) requires the client to choose one of six main options. Option A: priced contract with activity schedules is the preferred approach when the client can provide the contractor with a complete description of what is required at the outset, so that the contractor can price it with a reasonable degree of certainty. This does not necessarily mean a complete design, but could be a full and unambiguous statement of what is wanted, for example, as performance specification or scope design and a statement of the purpose of the asset.

Broome (1999) states that BofQs are the normal payment mechanism for work procured under the traditional/sequential procurement method, where the works are substantially designed by or on behalf of the employer before being put out to tender by the contractors. However, BofQs have some fundamental flaws, the biggest being that construction costs are rarely directly proportional to quantity. The effect of this flaw may be minimal if changes in the scope of the works are few, as the difference between the contractor's costs and income will remain small.

Table 12.4 Activity schedule for earth-fill dam for new reservoir (part only).

Ref	Activity	Price (£)
A001	Mobilize excavation plant	7,500
A002	Set up sit offices	25,000
A003	Site clearance	10,000
A004	Drainage to dam	45,000
A005	Divert steam	2,000
A006	Excavate abutment and inlet	20,000
A007	Excavate culvert	15,000
A008	Excavate discharge channel	5,000
A009	Excavate stilling basin	15,000
A010	Excavate spillway	20,000
A011	Excavate cascade channel	30,000
A012	Place fill to dam (core)	50,000
A013	Place fill to dam (clay fill – phase 1)	60,000
A014	Place fill to dam (clay fill – phase 2)	60,000
A015	Place fill to dam (clay fill – phase 3)	60,000
A016	Place loam and dress	30,000
A017	Construct road on top of dam	30,000
A018	Reinstate quarry	2,000

Use of method-related charges, separating fixed and time-related charges from quantity-related costs, increases the change in scope that can be accommodated. However, friction often results when significant changes in scope or methods occur, as the contractor struggles to justify any additional entitlement and to show where his extra costs come from. The situation is not helped by the lack of programming provision in traditional conditions of contract.

Clause 5 in the JCT98 standard building contract makes the provision of a contractor's programme an option. In contrast, clause 14 of the ICE7th civil engineering contract requires the contractor to submit a construction programme and method statement 28 days after the award of the contract; neither of the standard forms dictate the type or content of the programme.

Activity schedules are an attempt to move away from the problems associated with BofQs. In concept an activity schedule is similar to a series of bars on a bar chart (Gantt Chart). The difference is that each bar/activity has a price attached to it and the contractor is paid for each completed activity at the assessment date following its completion. The activity schedule is therefore closely linked to the programme and, as the contractor prepares the construction programme, the contractor would normally prepare the activity schedule. He would then also know his expected cash flow.

The value of the critical-path method, especially when it is developed into a full network analysis has long been recognized on major projects as an increasingly important tool in the control of the project. Dr Eric Cole's forecast in *The Civil Engineering Surveyor* in 1991 that network analysis could replace the traditional BofQ is moving a step closer (Cole, 1991). Broome's research found that the theoretical advantages, which the use of activity schedules should give, appeared to materialize in practice. Broome listed the advantages and possible disadvantages of activity schedules which are as follows.

General advantages

- Under performance-based contracts any significant level of contractor design is more easily accommodated, as design itself can become an activity.
- In order to receive payment as planned, the contractor has to complete an activity by the assessment date. Consequently he has to programme realistically and is motivated to keep

to that programme during construction. Throughout the contract the activity should mesh with the time, method and resource document (the NEC-accepted programme).

- Assessment of the amounts due to a contractor with an activity schedule is easier and involves many less man-hours than with a BofQ.
- As payment is linked to completion of an activity or group of activities, the cash-flow requirements for both parties are more visible.
- Contractors are not paid for changes in the quantity of permanent work, unless an instruction changing the original specification is issued, thus transferring the risk to the contractor.
- The assessment of the effect of a compensation event is easier with activity schedules than with BofQs.

Potential disadvantages

- Contractors have to plan the job before they prepare their activity schedule because they start from a blank sheet, rather than being given a detailed BofQ. Contractors are forced to prepare a more thorough tender; this, however, is a time-consuming task which increases the man-hours needed to prepare a tender and is ultimately charged to the employer. Employers may thus wish to put potential work out to fewer tenderers;
- While the assessment is easier and fairer, because any change in resources or methods associated with an activity can be compared with those in the accepted programme before the compensation event occurred, it is a more rigorous process than using bill rates and therefore takes longer. For this reason, both parties may be happy to use the bill rate for the assessment of a simple compensation event. However, this approach ignores the delay and disruption costs associated with any change which research has shown costs, on average, approximately twice the direct costs.

Broome found that because of the practical advantages there had been a definite shift particularly by experienced users of the NEC Contract away from Option B: (BofQ) to Option A: (activity schedule). However, where a contractor's costs are much more quantity related (e.g. in some building work) and/or variations in quantity are expected (e.g. in refurbishment projects), a method-related BofQ approach should be more appropriate.

Klein (1995) agreed that the activity schedule found within the NEC provided a better approach for fair payment. He considered that the NEC schedules include items of work in smaller packages than the amounts normally comprised within milestone payments, so there should be fewer arguments over whether the activity has been completed. But delays caused by the particular or specialist remains a problem.

12.8   Cost-based, cost-reimbursement contracts

Cost-reimbursement contracts, particularly when incentives are incorporated, have many advantages both for client and contractor. These include flexibility to change, fair apportionment of risk, potential saving in time and cost of tendering, open-book accounting, and a reduction in resources expended on claims (Perry and Thompson, 1982). Under cost-reimbursement contracts the contractor is paid a fee for overheads and profit based on the actual cost of construction. These types of contracts can provide a valid alternative to the traditional approach in certain circumstances as follows:

- where the risk analysis has shown that the risks are unconventional in nature or in magnitude;

- where the client is unable to define clearly the works at tender stage, substantial variations are anticipated, or there is an emphasis on early completion;
- where an increased involvement of the client and/or contractor is required or desirable;
- where there exists exceptional complexity, for example in multi-contract projects or where a high degree of technical innovation is demanded;
- emergency situations where time is of the essence;
- where new technologies or techniques are involved;
- where there is already an excellent relationship between client and contractor, for example, where there is trust.

Generally cost-reimbursement contracts eliminate a large number of risks in the project for the contractor and place them with the client. It is recognized that the client gains greater flexibility in a cost-reimbursement contract and the contractor can be confident of an equitable payment for changed and unforeseen events.

Cost-reimbursement contracts allow contractors early involvement at the design stage and allow client participation in the contract management. This approach identifies with the philosophy that it is cheaper in the long run for the employer to pay for what does happen rather than for what the contractor thought might happen in those areas of doubt, which the contractor cannot influence.

12.9   Cost-based, target-cost contracts

Incentives

Traditionally very few construction contracts within the UK have included positive incentives for performance; most rely on damages for non-performance, for example, liquidated damages, maintenance periods, retentions, bonds and warranties. By contrast, in the US, it not unusual to find construction contracts with some form of positive incentive.

In recent times the first incentive contracts in the UK were the so-called *lane rental contracts* for motorway reconstruction. Under these contracts, a contractor is offered a bonus to complete works early, but is charged for each day it overruns the contract period. Under this arrangement contractors have consistently demonstrated that they are able to improve on the Ministry of Transport's minimum performance standards.

Target contracts require a different approach when compared to traditional contracts. Target contracts demand that the promoter, contractor and project manager are all involved in the management and joint planning of the contract. The promoter furthermore is involved directly in the costing and influencing decisions made on risk.

Thus a target-cost contract when combined with a partnering/alliance approach enables the supply chain to be fully involved in the design and planning stages. Target contracts have two main characteristics: they are cost-reimbursement contracts and positive incentives are employed. A target-cost contract is thus a means of sharing the risk between the employer and contractor with the latter being encouraged to maximize his performance. Similarly, over-expenditure is usually apportioned so that the contractor suffers most by receiving a reduction in the fee covering head office costs and profit.

The concept of bringing all the parties together early and managing the costs during the design development was embraced by the Ministry of Defence in their two pilot prime contracting projects at Aldershot and Wattisham known as *Building Down Barriers* initiative. The contractor's progress can be monitored against three targets: performance, time and cost, although combinations of different categories can be applied.

Panel 12.1 Case study: East River tunnels, New York (1904)

The concept of target cost is not new. In 1904, The Pennsylvania, New York and Long Island Railroad Company wished to link up its Long Island system with Manhattan Island and its main terminal on 7th Avenue.

This major tunnelling project required four tunnels each 23 feet in diameter to be constructed under the East river. The job was anticipated to be difficult and dangerous – a real high-risk project. The client came to a private agreement with Yorkshire contractor Weetman Pearson without putting out to public tender, indeed no other contractors were interested. The agreement specified a payment of 18 million gold dollars (£3.5 million) with a project duration of five years. Pearson agreed to pay half of any excess cost up to $2 million and to take half of any savings as a bonus. He was thus backing his judgement up to $1 million or £200,000.

Eight tunnelling shields were required, for the ground conditions varied from quicksand requiring compressed air working, to hard rock requiring drilling and blasting. Despite tremendous difficulties including fires in the tunnels, the last of the tunnels were joined up three and a half years from the sinking of the first shaft.

Source: Young, 1966.

Table 12.5 Traditional construction costing and target costing compared.

Traditional construction costing	Target costing
Costs determine price	Price determines costs
Performance, quality and profit (and more rarely waste and inefficiency) are the focus of cost reduction	Design is the key to cost reduction, with costs managed out before they are incurred
Cost reduction is not customer driven, nor project/design team driven. It is driven by separate 'commercial' people	Customer input guides identification of cost-reduction areas
Quantity surveyors advise on cost reduction	Cross-functional teams manage costs
Suppliers involved late in design process	Early involvement of suppliers
No focus on through-life cost	minimizes cost of ownership for client and producer
Supply chain only required to cut costs – regardless of how it is done	Involves supply chain in cost planning

Source: Holti et al., 2000.

Performance targets

Performance targets have been used in the process-plant industry, particularly where the contractor is responsible for design. The contractor either earns a bonus or incurs a penalty that adds or subtracts from his earned fee, or earns an *award* fee, which is added to a minimum or base fee.

The performance is measured against the parameter, which has the most significant influence on construction cost and programme, for example, quality, safety, technical management and utilization of resources. However, as the payment is invariably based on client's subjective judgement of a contractor's performance, incentive adjustments made by the client are likely to be disputed by the contractor.

Time targets

Time targets operate in much the same way as for motorway reconstruction work; a bonus or penalty is awarded, depending on whether the contractor is ahead of, or behind, the construction programme. The bonus is generally a monetary amount per day and the penalty, an amount per day or loss of fee on work done past the completion date.

Cost targets

A typical cost-incentive system involves the sharing of target project-cost overrun and under run. Usually cost targets are set for the combined construction costs with an agreed separate fee for overheads and profit (Fig. 12.2). Other cost incentives which have been used more commonly in the process-plant industry, involve the sharing of target man-hour cost overrun and under run, or involve a bonus/penalty system based on the average man-hour cost.

The target price comprises four elements: the unit costs, the risk, the overheads/management fee and the profit. Depending on the amount of information available the target cost for a building project could be established using cost per unit (e.g. car park), cost per m^2 of floor area, built up on an elemental basis or based on detailed BofQ.

Targets should not be fixed until the design is 40–60% complete. The most common method of target setting appears to involve the use of a crude-priced BofQ (reflecting the major cost-significant items – Pareto rule: 80% cost in 20% of the items). However, in order for the

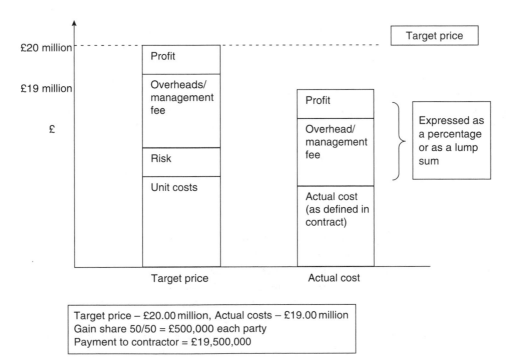

12.2 Calculation of gain share.

incentive to be maintained the target cost must be adjusted for changes in the scope of the work and matters outside the control of the contractor, for example, major variations and inflation. The *Engineering and Construction Contract* (NEC3) identifies 19 compensation events; under this contract all compensation events give rise to an extension of time and adjustment to the target.

Relevant observations

Renzo Piano, one of the most eminent architects of our time and joint designer with Richard Rogers of the iconic Pompidou Centre in Paris, makes a relevant observation concerning the new Pierpont Morgan Library in New York:

> I hate the idea that the architect, because he is an artist (which he is), may forget about little duties, such as money and function. To be on budget is not a miracle. It's the result of a lot of work and fighting. We have done 4,000 drawings for this job. I am the son of a builder. I love builders but I also know they are sharks. Well for them it is a game.
>
> (Binney, 2005)

Many would have agreed with these sentiments in the pre-Latham era; however, since then there has been a dramatic change in a significant part of the industry. Central and local government and some commercial clients are now embracing the new philosophies and strategies of partnering/alliances and PFI/PPP relationships which require openness and trust. This post-Egan era approach requires a radical new way of thinking and understanding of the philosophy and cultural aspects, the legal and financial frameworks that underpin the design, construction and operation process. The new approaches demonstrate a rejection of the short-term, potentially adversarial models in favour of a more transparent, longer-term and collaborative ways of working. However, we know that many of those who are involved in these approaches have yet to embrace the true philosophy of creating a win-win scenario for all the participants.

For example, how many PFI consortia currently building the nation's hospitals can honestly state that they are not knowingly designing obsolescence into the schemes? This sceptical view is confirmed by an alarming survey conducted by the Chartered Institute of Building to over 1,400 senior construction professionals. The results of the survey identified that the corruption is still prevalent in the UK construction industry; 60% of the correspondents felt that fraud within the industry was prevalent and 41% had been personally offered a bribe. Furthermore there was a concerning level of people who thought, for example, that producing a fraudulent invoice was not corrupt or that using a bribe to obtain a contract was also not a particularly corrupt practice. (www.ciob.org.uk/resources/research [accessed 3 February 2007]).

12.10   Conclusion

It is observed that the selection of a suitable payment system is dependent on the nature of the works and the degree of completion of the design.

Where high certainty of cost is desired, price-based systems should be useful not only for tender evaluation and interim payments but also for valuing variations and claims. Price-based systems are more suitable for design-completed projects of which design is completed or insignificant changes are envisaged. Lump-sum contracts are particularly suitable for completed design. In the event that a client wishes to retain the flexibility in making changes in the design, or ground conditions are uncertain, admeasurement contracts should be appropriate.

The common characteristics of the various cost-based systems is the high uncertainty in actual cost. Risk sharing, with a high degree of trust between the parties is the essence in a cost-based system.

We have also identified that the traditional BofQs have many shortcomings – not least that they expose the client to the risk of errors in them. The lump-sum plan and specification approach also has limitations and is only considered suitable for smaller projects. In comparison priced-activity schedules have many advantages particularly when linked to programmes through powerful computer software.

12.11   Questions

1.   On a multi-million pound inter-disciplinary project based on a BofQ, the client has suggested that, to ease the task of interim valuations, payment should be made based on overall percentage progress against the construction programme. Discuss the implications, advantages and/or disadvantages of this suggestion (RICS, Direct Membership Examination, 1985, Project Cost Management Paper).
2.   The use of BofQs as a contract document may be considered outmoded practice. Discuss this point of view (CIOB Member Part II, Contract Administration, 1987).
3.   Compare and contrast the use of priced BofQ with priced-activity schedules under the following headings.

   a.   For the client:
      i.   budget estimating/cost planning;
      ii. selecting contractor;
      iii. interim payments;
      iv. valuing variations.

   b   For the contractor:
      v.   estimating;
      vi. purchasing;
      vii. programming;
      viii. site management;
      ix. financial control.

Bibliography

Barnes, N.M.L. and Thompson, P.A. (1971) *Civil Engineering Bills of Quantities*, CIRIA Report 34, CIRIA

Binney, M. (2005) 'Notebook architecture', *The Times*, Monday, 17 October

Broome, J. (1999) *The NEC Engineering and Construction Contract: A User's Guide*, Thomas Telford

Broome, J. (2001) 'Activity schedules v bills of quantities', *The NEC Users' Group Newsletter*, issue no. 22, p. 8

Building Research Station (1968) 'Tendering documents with a production bias', *Digest* 97 (Second series), London, HMSO

Carrick, D. (1991) 'BofQs and networks exist,' *Civil Engineering Surveyor*, September, pp. 25–26

Cole, E. (1991) 'The future for quantity surveying', *Civil Engineering Surveyor*, May, pp. 10, 17

Gryner, D.I.B. (1995) 'The Linking of Programme and Payment through Milestone Contracts', First International Conference on Construction Project Management, Singapore, January

Hoare, D.J. and Broome, J. (2001) 'Bills of quantities v activity schedules for civil engineering projects', *Journal of Construction Procurement*, vol. 7, no. 1, pp. 11–26, May

Holti, R., Nicolini, D. and Smalley, M. (2000) *The Handbook of Supply Chain Management: The Essentials Building Down Barriers*, C546, CIRIA

Klein, R. (1995) 'Payments at fair stages', *Building* magazine, 17 March, p. 38

Mitchell, B. (2003) *Jubilee Line Extension: From Concept to Completion*, Thomas Telford

NEC (1995) *New Engineering and Construction Contract*, 2nd edition Guidance Notes, Thomas Telford

Perry, J.G. and Thompson, P.A. (1982) *Target and Cost-reimbursable Construction Contracts*, CIRIA Report R85, CIRIA

Potts, K.F. (1987) 'An alternative payment system for major "fast track" construction projects', *Construction Management and Economics*, vol. 6, no. 1, 25–33

Potts, K.F. and Toomey, D. (1994) 'East and West Compared: A Critical Review of Two Alternative Interim Payment Systems as Used in Hong Kong and the UK on Major Construction Works', CIBW92 Procurement Systems Symposium, 4–7 December, Hong Kong

Ridout, G. (1982) 'Target cost takes the risk out of contracting', *Contract Journal*, 14 October, pp. 16–18

Russell, P.J. (1994) 'Hong Kong's airport railway – part 2', *Civil Engineer Surveyor*, Dec/Jan, pp. 15–17

Seeley, I.H. (2001) *Quantity Surveying Practice*, Macmillan

Sharpe, D.J. (1987) 'Completing on Time – Construction Management', Conference, Urban Railways and the Civil Engineer, ICE, London, September/October, pp. 165–180

Skinner, D.W.H. (1981) 'The Contractor's Use of Bill of Quantities', Occasional Paper No. 24, CIOB

Skoyles E.R. (1968) 'Introducing bills of quantities (operational format)', *Quantity Surveyor*, vol. 24, no. 6, pp. 139–146

Skoyles, E.R. and Lear, R.F. (1968) 'Practical applications of operational bills 2, use of bill when employment of main contractor is determined', *The Chartered Surveyor*, vol. 101, no. 2, pp. 70–76

Wallace, I.N.D. (1986) *Construction Contracts: Principles and Policies in Tort and Contract*, Sweet and Maxwell

Young, D. (1966) *A Biography of Weetman Pearson, First Viscount Cowdray, Member for Mexico*, Cassell

Part V
Management of the post-contract stage

13 Contractors' cost-control and monitoring procedures

13.1 Introduction

Contractors generate small percentage of profit margins, between 2 and 7%, from large annual turnovers. For example the 2006 annual report of the UK's largest contractor Balfour Beatty indicates an after-tax profit of £125 million on an annual turnover of £5,852 million, that is, 2.13% (www.balfourbeatty.com [accessed 20 April 2007]). Without an effective company-wide cost-control system this would result in substantial risk to any contractor's slender margins.

Cost management is very much more than simply maintaining records of expenditure and issuing cost reports. Management means control, so cost management means understanding how and why costs occur and promptly taking the necessary response in light of all the relevant information. Keeping a project within budget depends on the application of an efficient and effective system of cost control. From the information generated it should be possible not only to identify past trends but also forecast the likely consequence of future decisions including final out-turn cost, that is, the final account.

Bennett (2003) identifies that there are three purposes of a contractors' cost-control system:

1. to provide a means of comparing actual with budgeted expenses and thus draw attention, in a timely manner, to operations that are deviating from the project budget;
2. to develop a database of productivity and cost-performance data for use in estimating the costs of subsequent projects;
3. to generate data for valuing variations and changes to the contract and potential claims for additional payments.

Two related outcomes are expected from the periodic monitoring of costs:

1. identification of any work items whose actual costs are exceeding their budgeted costs, with subsequent actions to try to bring those costs into conformance with the budget;
2. estimating the total cost of the project at completion, based on the cost record so far and expectations of the cost to complete unfinished items.

Barnes (1990) reinforces these concepts identifying two critical factors, which should be considered in the financial control of any construction project.

1. The methods of control of a project should be appropriate not only in its objectives and size, but also to the uncertainties inherent in predicting its cost, timing and risk of changes.

Uncertainty is inevitable on all projects and could include interest rates changes, client changes, low productivity, unforeseen ground conditions, failure of specialist contractors and the weather.

2. Control must include action. We cannot control the past so our effort and energy in managing any project should be focused on controlling the present and future and taking necessary corrective action. It will be necessary to consider alternative actions the consequences of which must be forecast.

Oberlender (1993) concurs identifying that cost control is far more than controlling expenditure. Cost control also includes the control of revenue, making sure that all possible and justifiable income is recovered from the client and that no preventable wastage of money or unauthorized increase in costs is allowed to happen.

13.2   Developing a cost-control system

The type and sophistication of any cost-control system will be determined by the resources available to operate the system and the use made of the system by the relevant management personnel. Pilcher (1994) considers that a wide variety of issues need to be considered when developing a contractor's cost-control system, namely the size of company, the type of work – building or civil engineering and the different contractual arrangements. He also notes the two main approaches and highlights the potential problems.

1. An integrated reporting system (integrating time and cost):
    Pilcher (1994) considered that integrated systems had the disadvantage that either simplicity or attention to level of detail was sacrificed. In other words this approach was good for understanding the big picture – was the project ahead/behind programme what was the internal cost what was the earned value, that is, the project variances?

An example of this approach is the 'earned value analysis' identified in BS 6079 *Project Management*. The big advantage of this approach is that one integrated system can be used for the control of time and cost. Nowadays, with the rapid development of sophisticated computerized databases, for example, ORACLE, it is possible to further analyse the project costs and identify the financial trends within the various sections or components of the project.

2. Separate schedule and cost-control systems:
    According to Pilcher (1994) it is the experience of many practitioners that separate schedule and cost-control systems provide a cheaper means of good control and the output from the two systems is more easily understood. However, he notes a word of caution that the processing of the data for separate systems does, of course, need to be integrated. This approach should enable the project team to identify the problem areas and take the necessary action – if possible.

Harris and McCaffer (2006) recommend that cost control should be exercised before the costs are committed. They point out that most cost-control systems have an inordinately long response time. Even the best current system provides information on what was happening last week or last month. Cornick and Osbon's (1994) research found that traditionally contractors' quantity surveyors (QS) only monitored costs rather than control costs which made their role reactive rather than proactive.

In conclusion, an effective cost-control system should contain the following characteristics:

- a budget for the project set with a contingency figure to be used at the discretion of the responsible manager;
- costs should be forecast before decisions are made to allow for the consideration of all possible courses of action;
- the cost-recording system should be cost-effective to operate;
- actual costs should be compared with forecasted costs at appropriate periods to ensure conformity with the budget and to allow for corrective action if necessary and if possible;
- actual costs should be subject to variance analysis to determine reasons for any deviation from the budget;
- the cost implications of time and quality should be incorporated into the decision-making process.

There are two ready-made types of contractor's cost-control systems:

1. the big-picture monthly review of the project;
2. a more detailed analysis of the sections within the project in order to identify those sections of the works which are underperforming.

The three main types of contractors' project cost-control systems are as follows:

1. cost-value reconciliation (used by building contractors);
2. contract variance – unit costing (used by civil engineering contractors) (see Fig. 13.1);
3. earned value analysis (US approach/used on major projects).

13.3   Method 1: cost-value reconciliation (CVR)

Cost-value reconciliation (CVR) brings together the established totals for cost and value to illustrate the profitability of a company. Its intention is to ensure that the profits shown in the company accounts are accurate and realistically display the current financial position.

The CVR serves two purposes; first, it forms the basis of statutory accounts, which is a legal requirement. The guidelines of *Standard Statement of Accounting Practice No. 9* (SSAP9), (ICAEW, 1998) states that 'valuation of stocks and works in progress' must be followed. The main thrust of SSAP9 is that financial statements should be prudent, that losses or potential losses should be recognized immediately and essentially that the business should not claim to be more profitable than it actually is.

Second, the CVR provides management information to assist in the identification of problems, the need for reserves, the reasons for loss and information to prevent repetition of such losses. It should also show the original budget figures and expected profit together with an assessment of the final position of the project, that is, the final account.

At each interim valuation date, normally the end of each calendar month, the total costs to date are compared with the total valuation. Care has to be taken to compare like with like and make necessary adjustments for overvaluation/undervaluation. This approach suffers from the disadvantage that there is no breakdown of the cost/profit figures between the types of work or different locations within the project; it therefore only provides guidance on which project requires senior management attention. This approach is suitable for use on building projects where there are a large number of complex components.

The Chartered Institute of Building (CIOB) *Cost Valuation Reconciliation* approach (described in detail in Barrett, 1992) is the standard recommended approach, a similar version of which is used by most contractors.

The cost-value comparison, or reconciliation, is usually completed by the contractor's QS on a monthly basis following agreement of the interim valuation. The process will require liaison with other departments in its completion and considerable discussion with the rest of the project team for example contracts manager and site manager.

The starting point of any CVR must always be the gross certified value, which must be supported by the architect's interim certificate. This is the external valuation and not the contractor's QS's assessment or internal valuation of the works. It is generally necessary to adjust the external gross certified value for sundry invoices, that is, work carried out on or off the project; using labour/plant/materials/subcontractors but which do not form part of the contract works.

An external valuation may require adjustments for many reasons, not least arithmetical errors found in the external valuation after agreement with the client's QS. Common areas of adjustment are as follows:

- adjustments for external preliminaries claimed in valuations against the internal preliminary schedule;
- adjustments for elements included within costing but not in the external valuation, for example, materials brought to site on the same day as the valuation but after materials on site were recorded;
- items of over measurement not picked up by the client's QS;
- any adjustments necessary to bring the cost cut-off date and the on-site valuation date together;
- variations which have not been agreed with the client's QS;
- contractual claims for loss and/or expense which have not been agreed by the client's QS;
- possibility of liquidated damages being charged by the employer;
- provision should be made for future known losses.

The general principle for the contractor's QS to remember must be that of caution. Any figures included in the valuation adjustment must be capable of substantiation and wherever possible have been agreed with the client's QS.

Once the gross adjusted valuation has been assessed three further elements are deducted from the figure to arrive at a final residual value or margin.

First, subcontract liabilities which should cover all disciplines of subcontractor, that is, nominated/named or domestic subcontractors, but excluding labour-only subcontractors, the cost of which should be considered with labour section of the main contractor's costing.

The subcontract liability is essentially a comparison between what the contractor has been paid and what the contractor is liable to pay to any given subcontractor. The liability should include not only those matters listed on the external valuation, such as contract works, variations and materials on site but also any works that the main contractor is due to pay the subcontractor but for which he would not receive reimbursement through the main works contract provisions.

Second, snagging and defects which can be subdivided into two sections:

- Snagging required at the end of the project to achieve handover;
- A levy to be used to cover for costs that may be incurred in the making good of defects period before the certificate of making good of defects is issued.

Both these figures are highly dependent on the type of project involved. On housing projects, an allowance per unit may be adopted. Many contractors will have standard allowances, for example, 1.5% of the contractors' gross value (excluding subcontract figures) building up over the contract period, and then reducing to 0.5% after practical completion is achieved.

The third and final element of deduction is that of the *main contractor's core costs*, that is, labour, material and plant and other associated costs. In general, these will be supplied to the contractor's QS by a separate cost department within the contractor's company.

Once all these figures are known the residue is the profit generated from the main contractor's works section of the project. When added to the profit on other sections, that is, the subcontract liability schedule, this indicates the profit to date for the project. The figure can then be compared with the original contract profit included in the contractor's tender.

In order to complete the *time analysis* (in weeks) section of the report it will be necessary to consider the following: the original period, any extension of time (EoT) awards, the present position and time to complete. If the contract period will be exceeded the contractor should include for: the cost of *preliminaries* in overrun period, any liquidated and ascertained damages, under recovery of fluctuations and claims from subcontractors and suppliers.

There is no room in a cost-value reconciliation for historical costing only. Without a clear vision of the completed project financial reporting at an intermediate stage of the project, it does not produce accurate profit and loss statements.

The preparation of cost-value reconciliations is not an exact science, particularly when completing Section 4: Provisions; these figures are estimated based on the QS's best current knowledge and experience. However it is important that there is a consistent approach taken by all the company's Quantity Surveyors. The author recollects that the Chief Quantity Surveyor of Holland Hannen & Cubitts (North West), who were later taken over by Carillion, held regular communication meetings with all the company Quantity Surveyors present. It was at these meetings that the monthly cost reports were discussed in detail and a standard company policy established.

Research by Stephenson and Hill (2005) noted that some contractors, mainly the larger ones, proposed the application of CVR in conjunction with an IT-based budget-monitoring system (BMS). A sophisticated BMS enables the incurred costs to be linked to the progress as shown on the programme. Costs to complete and the anticipated financial out-turn costs can also be established.

More recent research by Lee Chun Hoong (2007), which was based on a return of 33% of the top-100 UK contractors, identified that the most popular cost-control system was the monthly CVR approach (58%) followed by standard costing by sub-variances (18%). The time-frame for the final preparation of the CVR was 14 days after the valuation date (73%). This research also showed that cost-control systems were used to compare the actual expenses versus the budget (100%); to forecast project completion cost and potential profit or loss for the project (100%); to analyse performances of the project (actual progress versus planned programme) (89%); to generate data for valuing variations (32%); to notify potential future claims (32%) and as basis for the valuation for subcontractors' interim payments (32%).

EXAMPLE OF MONTHLY CVR

Prepare the contractor's cost-value reconciliation (internal valuation) for work done up to the end of November 2008 on the following project:
High-street development with ground-floor shops and three-storey offices above.

Contract particulars: JCT 2005 *Private with Quantities*
Fluctuations: Schedule 7 Option A applies (only changes in
                              contributions, levy and tax fluctuations reimbursed)
Accepted tender: £2,555,000
Contract period: 15 months
Date of commencement: 1 January 2008

Particulars appertaining at end of November 2008:

Cost to end of November 2008   £1,850,000
Contractor's interim valuation – gross total   £2,055,000
Claims disputed by architect/QS: £255,000
Separate invoice to employer: £7,500 (building a garage at
                                                                client's home)

Undervalue of materials on site: £10,000
Overvalue on preliminaries: £7,750
Overmeasure: £17,850

Other issues:

- M&E s/c's application only includes for work up to 20 November 2008 (main contractor's application includes an extra £15,000);
- vandalism/theft likely;
- higher-than-anticipated rate of inflation;
- project running three weeks behind master programme (liquidated damages £5,000 per week);
- One-week EoT awarded (due to late issue of instructions).

Solution: At first sight this project would seem to be showing a profit of £205,000, with an interim valuation of £2,055,000 against an internal cost of £1,850,000. However further analysis is required in order to establish a more prudent assessment in accordance with SSAP9 standard accountancy practice. We need to ensure that we are comparing like with like (valuation and costs) and also make further allowances for anticipated problems.

Valuation

1. *Application and certificate*

		£	
	(a) Total payment application (net of discount and VAT)	2,055,000	
Deduct	(b) Disputed items, claims not agreed in application	(255,000)	
	(c) Anticipated gross certificate	1,800,000	1,800,000
Add	(d) Additional agreed contract invoices	7,500	7,500

2. *Adjustments*

Add (a) Subcontractors' adjustments   —
     (b) Preliminaries   —

(c) Valuation before close of accounting period   —
(d) Materials on site   10,000
(e) Variations not agreed   —

        10,000   10,000

3. *Overvaluation*

Deduct: (a) *Preliminaries*   7,750
        (b) Valuation after end of month   —
        (c) Weighted items   —
        (d) Materials on site   —
        (e) Overmeasure   17,850
        (f) M&E subcontract   15,000

        40,600   (40,600)

4. *Provisions*

Deduct (a) *Remedial works*   5,000
       (b) Winter working   5,000
       (c) Foreseeable risks   —
       (d) Unprofitable future work   —
       (e) Defects liability period   2,000
       (f) Shortfall in increased costs recovery   5,000
       (g) Vandalism/theft   5,000
       (h) Late completion:
          Liquidated damages two weeks at   £5,000   10,000
          Preliminaries two weeks at £3,850   7,700
          Increased costs two weeks at £500   1,100
          Claims from subcontractors   5,000   23,700

       (45,700) (45,700)

5. *Internal valuation*

        1,731,200

Summary

So after making the necessary adjustments a more realistic financial comparison emerges. This shows that the project is actually making a loss of £118,000 (adjusted internal valuation of £1,731,200 against an internal cost of £1,850,000).

13.4   Method 2: contract variance – unit costing

In this system costs of various types of work, such as driving piles, or concrete work are recorded separately. The actual costs are divided by the quantity of work of each type that has been done. This provides unit costs, which can be compared with those in the tender.

   The report is prepared on a monthly basis following the interim valuation agreed with the client. The report requires a comparison to be made between the value of the work done and

Mighty Build Construction Company Monthly cost report Programme as tender: 100 weeks
Contract: Merrythorpe Marina, N.E. England Date: end of month To date: 75 weeks; Actual: 65 weeks

Cost code	Description of work	Unit	BofQ	To date	Quantity estimated final Total	To complete	BofQ	Cost to date	Valuation to date	Cost +/- to date	(£1,000) Estimated final cost	Final valuation	Final +/-
010	General items	Item	Fixed	90%		10%	500	500	450	(50)	556	500	(56)
		Weeks	Time related	75%		25%	1,000	800	650	(150)	1,354	1,000	(354)
020	Excavate over site and remove	M³	450,000	500,000	500,000	Complete	2,250	2,200	2,500	300	2,200	2,500	300
030	Concrete piling to marina walls	Nr.	250	255	255	Complete	1,000	950	1,020	70	950	1,020	70
040	Marina excavation and remove	M³	700,000	350,000	700,000	350,000	4,900	2,800	2,450	(350)	5,600	4,900	(700)
050	Lock construction cofferdam	M²	5,000	2,000	5,000	3,000	1,250	650	500	(150)	1,625	1,250	(375)
060	Lock construction gates	Nr.	2		2	2	500						
070	Piling to pontoons	Nr.	400		405	405	800						
080	Puddle clay bed	M³	100,000		100,000	100,000	1,000						
090	Flood marina	M³	600,000		600,000	600,000	600						
100	Install pontoons	M	2,000		2,000	2,000	1,000						
110	Finishing work	Item				100	100						
							14,900	7,900	7,670	(330)	12,285	11,170	(1,115)

13.1 Monthly cost report based on the Contract Variance Unit Costing Method.

the cost of doing it, that is, the variance. The aim of the report is to identify the problem areas and trends as well as forecasting the final profit/loss situation on the project. Corrective action should be taken on any cost centres showing a loss if at all possible. This form of cost report is most effective on a contract with repetitive operations but is less so on non-repetitive contracts. This approach is appropriate for civil engineering work where there are a small number of high-value components in the project.

EXAMPLE OF MONTHLY UNIT COSTING METHOD

This case study relates to the construction of a new marina in Merrythorpe on the coast of the northeast of England. In the mid-1930s the site was used for gravel extraction. These operations ceased in 1939 and after this the site was used as a tip. The tipping ceased in the late-1950s and in the early-1960s the local authority constructed an incineration plant which was revamped with increased capacity in the late 1990s.

The time for completion of the project is 100 weeks with a contract value of £14.9 million. The cost report (see Fig 13.1) reflects the financial situation at the end of week 75 and it is noted that the project is 10 weeks behind the tender programme, the finishing date being week 110.

The general items or preliminaries have been split into two items representing:

- *fixed charges*: cost of mobilization and demobilization of construction equipment together with the contractors' facilities;
- *time-related charges*: representing the weekly cost of running the site include costs of all staff, maintaining the construction equipment and fuel.

Of the *fixed charges* within the general items, 90% has been paid to date (£450,000) and this is considered a realistic assessment of the situation; the actual costs however are shown as (£500,000), thus showing a loss to date of (£50,000).

All the estimated final costs for each cost code have been calculated pro rata to the actual costs to date. The final valuation figure is based on the initial tender figure adjusted if necessary for any variations, that is, both the final cost and valuation are calculated representing the final estimated quantity.

The client's contract administrator has paid for only 65 weeks *of time-related* costs – reflecting the actual progress to date. The report shows a final valuation of £1,000,000 (as tender), with the final costs based on 110 weeks that is £1,354,000 – an anticipated final deficit of £354,000.

Overall the project is showing an anticipated loss of £1,115,000. Two major issues have occurred on this project that have affected progress. First, more contaminated material has been encountered than was envisaged from the bore-hole reports available at tender; this has caused a six-week delay. Second, whilst constructing the lock cofferdam granite boulders were encountered. In order to overcome this latter problem the contractor had to devise a revised working method, which included drilling and blasting the rock, resulting in an additional delay of four weeks.

Under a traditional contract (e.g. ICE or FIDIC) these two items would be dealt with as claims for both costs and EoT under the unforeseen ground conditions clause in the conditions of contract (e.g. ICE clauses 12 and 44). These two items would obviously have a significant impact on the contractor's cash flow and the contractor's commercial team would be under considerable pressure to evaluate and justify the claims submitted. The contract variance cost reporting approach should allow the contractor to establish the broad financial figures involved.

If the contract was let based on an NEC ECC contract the two items would be considered compensation events and the contractor would be required to submit quotations for the work. If the quotations were accepted it is suggested that the valuation figures shown on the monthly cost report should include the accepted amounts.

If the valuation was the same but the estimated costs (in £1000s) to date (on week 65) were as follows:

010	General items: fixed	425
010	General items: time-related	700
020	Excavate over site and remove	2,100
030	Concrete piling to marina walls	900
040	Marina excavation and remove	2,150
050	Lock construction cofferdam	350
060	Lock construction gates	—
070	Piling to pontoons	—
080	Puddle clay bed	—
090	Flood marina	—
100	Install pontoons	—
110	Finishing work	—

Identify the final estimated contract variance.

13.5   Method 3: earned value analysis

Traditional earned value analysis (EVA) is defined by Howes (2000) as 'an established method for the evaluation and financial analysis of projects throughout their life cycle'. Earned value management (EVM) is a fully integrated project cost- and schedule-control system which allows through trend analysis, the formation of 'S' curves and cost/schedule variances.

The technique can be applied to the management of all capital projects in any industry, while employing any contracting approach. EVM is superior to independent schedule and cost control for evaluating work progress in order to identify potential schedule slippage and areas of budget overruns. In the US from 1997 onwards private industry have started to adopt the EVM technique as it represented a viable, best-practice tool that project managers could actually use (Fleming and Koppleman, 2000).

The concept of EVA was initially conceived at the start of the twentieth century by industrial engineers in an attempt to improve production methods. Later it was utilized by the US Air Force. The UK construction sector has been reluctant to embrace the concept of EVA. However it has been used with some success by the major players including WS Atkins on the Channel Tunnel and Tarmac Construction (now Carillion) on the £80 million widening of the M6 at the Thelwell Viaduct, near Manchester. The Thelwall Viaduct project was based on the traditional ICE 5th *Conditions of Contract* with bills of quantities (BofQ). Tarmac Construction spent £28,000 reconfiguring the costs in the BofQ into a work breakdown structure (WBS) database in order to employ the EVA approach.

In recent years other major users of the EVA include Taylor Woodrow Construction, Skanska UK (civil engineering), Balfour Beatty and Edmund Nuttall Ltd (Wiggin, 2005).

Calculating the earned value

EVM involves calculating three key values for each activity in the WBS:

1. *The planned value* (PV): formerly known as the *budgeted cost of work scheduled* (BCWS) – that portion of the approved cost estimate planned to be spent on the given activity during a given period;
2. *The actual cost* (AC): formerly known as the *actual cost of work performed* (ACWP) – the total of costs incurred in accomplishing work on the activity in a given period. The actual cost must correspond to whatever was budgeted for in the PV and earned value (EV) (e.g. all labour, materials, construction equipment and indirect costs).
3. *The earned value* (EV): formerly known as the *budget cost of work performed* (BCWP) – the value of the work actually completed.

These three values are combined to determine at that point in time whether or not work is being accomplished as planned. The most commonly used measures are the cost variance and the schedule variance:

Cost variance (CV) = EV−AC

Similarly the cost of impact of schedule slippage, the schedule variance in terms of cost, may be determined.

Schedule variance (SV) = EV−PV

The same data can be expressed as ratios that give an indication of value for money. If work is proceeding to, or better than plan, these ratios will be equal to or greater than 1.0. Conversely unfavourable variances will be less than 1.0.

1. How are we doing on money?
    Cost performance index (CPI) = EV/AC

2. How well are we doing on time?
    Schedule performance index (SPI) = EV/PV

The EVM approach provides a most powerful control tool. The data generated should enable senior management to identify the performance of the project as a whole, or within any part of the project, at any point in time. Furthermore monthly trends can be easily identified by comparing the monthly cost performance index (CPI) and schedule performance index (SPI) figures. In addition, the EVM approach enables the forecast of the out-turn situation.

13.6   Conclusion

Developing and operating effective contractors' cost-control systems is a major challenge to the commercial managers due to the unique nature of construction projects. The information collected often relates to the past and reflects the costs of fixed items of major construction

---

Panel 13.1 Case study: Skanska Civil Engineering – use of EVM on Channel Tunnel Rail Link section 2 contract

The Channel Tunnel Rail Link is a major element of the UK Government's PPP which enables important infrastructure to be provided for the benefit of the public sector, while taking advantage of the private sector management and efficiency.

In 1996 London and Continental Railways (LCR) was selected by the government to build and operate the Channel Tunnel Rail Link, and to own and run Eurostar (UK) Ltd, the UK arm of the Eurostar train service. LCR's shareholders are Bechtel Ltd, SG Warburg & Co Ltd, National Express Group plc, French Railways Ltd, Systra-Sofretu-Sofrerail, EDF Energy Ltd, Arup Group Ltd and Sir William Halcrow & Partners.

Skanska Civil Engineering executed four contracts with a value of more than £500 million on the £7 billion Channel Tunnel Rail Link. The project under consideration comprised two tunnels of open-box-type structure at Stratford and East Pancras. The procurement method was two-stage design and build utilizing an NEC–ECC target cost contract with the client's activity schedule.

The client (LCR) instructed all contractors to use an earned value management approach and required a standardized reporting system to be implemented.

The general approach to implementing EVM required:

- the production of a programme with milestone work packages;
- the production of a project/cost payment schedule that correlated with the project programme;
- a physical measure each month, of all the work undertaken within each of the works packages;
- a calculated cost of the work completed;
- inputting the information obtained into Primavera P3 to achieve the EV output.

Compared to the traditional approach no additional work was required to operate the system; however, there was an increase in understanding and communication between the commercial and project teams. It was possible within the system to embrace the substantial variations and delays; the cost projection of the final account was also accurate. Future developments include encouraging subcontractors to use the EVM approach thus benefiting the whole of the supply chain particularly when bidding for future projects.

Source: Wiggin, 2005.

---

equipment and temporary works which may prove impossible to alter even if they are showing a loss.

There needs to be a balance between the cost of developing and operating a system and the potential benefits. The more sophisticated systems are expensive to develop and operate and may not produce the information required, for example, to support quotations for variations or justify expenditure after the event.

The author recollects being involved in operating a highly detailed computerized contractor's cost-control system in which every BofQ item was analysed at tender into labour/materials/plant and then each invoice and cost incurred was allocated to these same cost centres as the work proceeded. The system became a monster which required feeding monthly with vast amounts of data including gross quantities of completed work and analysis of items not described in the BofQ.

With hindsight this system was far too ambitious on a complex civil engineering project. Indeed it deflected the commercial team from the main task of producing the final account, finalizing the subcontractors' accounts, valuing the changes and evaluating the claims. However, this highly detailed system would probably work well on a more straightforward project, for example, a housing development.

The rapid development of computer hardware and software has greatly assisted in the computerization of the control process with an integration between the accounting and estimating software while supporting the needs of a cost-control system.

13.7   Question

1. Critically review the effectiveness of your own company's cost-control system, how effective is it?

Bibliography

Ahuja, H.N. (1980) *Successful Construction Cost Control*, John Wiley & Sons

Barnes, M. (ed.) (1990) *Financial Control*, Engineering Management series, Thomas Telford

Barrett, F.R. (1992) *Cost Value Reconciliation*, 2nd edition, CIOB

Bennett, F.L. (2003) *The Management of Construction: A Project Life Cycle Approach*, Butterworth Heinemann

Cornick, T. and Osbon, K. (1994) 'A study of contractors quantity surveying practice during the construction process', *Journal of Construction Engineering and Management*, vol. 12, no. 2, pp. 107–111

Fleming, Q.W. and Koppleman, J.M. (2000) *Earned Value Project Management*, 2nd edition, Project Management Institute

Fleming, Q.W. and Koppleman, J. (2002) 'Using earned value management', *Cost Engineering*, vol. 44, no. 9, pp. 32–36, EBSCO Publishing

Harris, F. and McCaffer, R. (2006) *Modern Construction Management*, 6th edition, Blackwell Publishing

Hayes, H. (2002) 'Using earned value analysis to better manage projects', *Pharmaceutical Technology*, vol. 26, no. 2, pp. 80–84, Contract Services

Howes, R. (2000) 'Improving the performance of earned value analysis as a construction project management tool', *Engineering Construction and Architectural Management*, vol. 7, no. 4, pp. 399–411, Blackwell Science Ltd

Kim, E., Wells, W. and Duffey, M. (2003) A model for effective implementation of earned value management methodology, *International Journal of Project Management*, vol. 21, no. 5, pp. 375–382, Elsevier Science Ltd and IPMA

Lee Chun Hoong (2007) 'A Critical Examination of Contractors' Cost Control Systems in the UK', unpublished BSc (Hons) Commercial Management and Quantity Surveying dissertation, University of Wolverhampton.

Oberlender, G.D. (1993) *Project Management for Engineering and Construction*, McGraw-Hill

Pilcher, R. (1973) *Appraisal and Control of Project Costs*, McGraw-Hill

Pilcher, R. (1992) *Principles of Construction Management*, 3rd edition, McGraw-Hill

Pilcher, R. (1994) *Project Cost Control in Construction*, 2nd edition, Blackwell Scientific Publications

Raby, M. (2000) 'Project management via earned value', *Work Study*, vol. 49, no. 1, pp. 6–9, MCB University Press

Staffurth, C. (ed.) (1975) *Project Cost Control Using Networks*, 2nd edition, Heinemann

Stephenson, P. and Hill, M.S. (2005) 'Cost Value Reconciliation (CVR) in the UK Construction Industry', RICS COBRA Research Conference, Brisbane, Australia, 4–8 July

The Institute of Chartered Accountants in England and Wales (ICAEW) (1998) Standard Statement of Accounting Practice 9 (Revised): 'Stocks and Long-term Contracts', London, ICAEW

Walker, I. and Wilkie, R. (2002) *Commercial Management in Construction*, Blackwell Publishing

Wiggin, C.P. (2005) 'Critical Evaluation of the Use of Earned Value Management in the UK Construction Industry', unpublished MSc Construction Project Management dissertation, University of Wolverhampton

Useful website

http://en.wikipedia.org/wiki/Earned_value_management

14 Change management – valuing variations

14.1 Introduction

Variations are inevitable on building and civil engineering projects and may range from small changes having little consequential effects to major revisions, which result in considerable delay, and/or disruption to the project.

There are a number of reasons for the introduction of changes on building works including: inadequate briefing from the client, inconsistent and late instructions from the client, incomplete design, lack of meticulous planning at the design stage, lack of coordination of specialist design work and late clarification of complex details (Gray *et al.*, 1994). Additionally, on civil engineering works, there are many cases where changes and new rates are necessary because of the nature of the ground. Furthermore, changes may occur due to the client's desire to include the latest technology.

Establishing a realistic valuation for variations on construction works is often not an easy task. Both parties need considerable experience and sound judgement to settle variations. The parties are required to have a sound appreciation of the methods of construction, estimating practice and scheduling of construction works often utilizing computer planning software. But most importantly the parties should keep comprehensive meticulous records of the factors relevant to the variation.

Panel 14.1 Case study: London Underground Jubilee Line Extension

It was always the intention to have a full *engineer's design* for the civil works with full working drawings produced to form part of the tender package – albeit contractors were also encouraged to submit alternative design and construction proposals. The very tight timescales for the original design phase and the changing requirements meant this objective was ambitious and, in practice, could not be realized.

Consequently the working drawings issued at contract award, despite the moratorium, remained incomplete in terms of both number and substance. This was highlighted in one contract where the contractor stated that they had been issued with 48,000 instruments of change by the time the work was complete.

Source: Mitchell, B., 2003.

The traditional method of valuing variations, both on building and civil engineering works, is to base the valuation of the variations on the rates or prices contained within the BofQ or schedules or pro rata thereto and only in extreme conditions was a *fair valuation* made. Furthermore, it was general practice to value the variation based on 'the rate which the contractor would have inserted against that item had it been included at the time of tender' (Haswell, 1963).

This approach of valuing variations often led to disagreement between the parties with the client's quantity surveyor (QS) wishing to rigidly adhere to the rates in the bill and the contractor wanting the rates to reflect the true cost as incurred or likely to be incurred. An analysis of the *Building Law Reports* shows that the valuation of variations has been a popular topic of litigation within the UK construction industry.

If the varied works are complex, the parties need to be skilled negotiators and be prepared to adopt a give-and-take attitude in order to bring about a satisfactory settlement. Under the traditional approach, compromise was often required for there was seldom one correct solution. Indeed the parties might consider several different approaches before selecting the appropriate strategy.

However in recent years there has been a shift of approach with many standard conditions of contracts introducing the requirement for the contractor to submit a lump-sum quotation for the variation prior to receipt of the official variation order and before carrying out the work. The advantage to the employer in this approach is that the final commitment, including disruption and extended time, is known prior to the instruction and the majority of the risk is transferred to the contractor. The advantage to the contractor is the certainty of obtaining adequate recompense for the variation – provided the quotation covers the full amount of the uncertainty.

14.2   Contractual requirements – *ICE Conditions of Contract*, 7th edition, January 2003

This form is designed for use on civil engineering projects based on a bill of quantities (BofQ) measured in accordance with the *Civil Engineering Standard Method of Measurement*, 3rd edition produced in 1991. The quantities set out in the BofQ are the estimated quantities of the work. The standard form creates what is known as an *admeasurement* contract by which the employer undertakes to pay for the actual quantities of work executed reflecting the engineer's design of the permanent works calculated based on the latest drawings and schedules.

The valuation of variations under clause 52 in the ICE *Conditions of Contract* (7th edition) is as follows:

Option 1 – sub-clause 52(1): If requested by the engineer, the contractor should submit his/her quotation for any proposed variation and his/her estimate of any consequential delay which should be agreed *before* the order is issued or *before* work starts.

Option 2 – sub-clause 52(3): Where a request is not made or agreement is not reached under option 1 as soon as possible *after* receipt of the variation, the contractor should submit to the engineer his/her quotation for any extra or substituted works *having due regard to any rates or prices included in the contract* together with their estimate of the cost of any such delay.

The engineer then has 14 days from receiving the submissions to either accept or negotiate. Upon reaching agreement with the contractor, the contract price should be amended accordingly.

Option 3 – sub-clause 52(4): Failing agreement between the engineer and contractor under either sub-clause (1) or (2) the value of variations ordered by the engineer in accordance with clause 51 should be ascertained by the engineer after consultation with the contractor in accordance with the following two principles and should be notified to the contractor.

1. Where work is of a similar character and executed under similar conditions to work priced in the Bill of Quantities it shall be valued at such rates and prices contained therein as may be applicable.

2. Where work is not of a similar character or is not executed under similar conditions or is ordered in the Defects Correction Period the rates and prices in the Bill of Quantities shall be used as the basis for valuation so far as may be reasonable failing which a fair valuation shall be made.

Under sub-clause 52(6) the engineer could instruct that any additional or substituted work be carried out on a daywork basis – this should be minor or incidental work which cannot easily be measured. For further discussion on dayworks see sub-clause 13.6 discussed in Chapter 17, *FIDIC Contract*, of this book.

The 2003 edition of the ICE 7th introduced the new sub-clause 52(2) which refers to sub-clause 51(3) *variations proposed by the contractor*, in other words savings initiated through value engineering.

Furst and Ramsey (2001) note in Keating that the procedure under sub-clause 52(3) falls short of the 'Change Order Procedure found under many contracts based on the U.S. practice, where the change order may not be issued until agreement is reached on price and delay'.

14.3 Contractual requirements – JCT *Standard Building Contract with Quantities* (SBC/Q 2005 edition)

At the instigation of the architect/contract administrator under clause 5.3.1, a quotation may be offered by the contractor for the work in accordance with schedule 2 quotation. The schedule identifies that the quotation should include the following items with sufficient supporting information:

1. Amount of adjustment to the contract sum – which should be made by reference to the contract bills, where appropriate, and with appropriate adjustment to the preliminary items;
2. Any adjustment in time required for completion;
3. Amount to be paid for direct loss and/or expense not included elsewhere;
4. A fair amount for the cost of preparing the schedule 2 quotation;
5. Where required by the instruction information on additional resources required to carry out the variation and the method of carrying out the variation.

If the employer wishes to accept the schedule 2 quotation, then the architect/contract administrator should confirm the quotation and issue a variation to the contractor making any necessary adjustment to the contract sum and the completion date.

If the employer does not accept the schedule 2 quotation, then the variation is valued based on the traditional 'valuation rules' contained within clauses 5.6 to 5.10.

Clause 5.6 also requires that if the work can be properly valued by measurement, then such work shall be valued in accordance with the following rules:

1. Where the additional or substituted work of similar character is executed under similar conditions as, and does not significantly change the quantity of, work set out in the Contract Bills, the rates and prices for the work so set out shall determine the valuation;

2. Where the additional or substituted work is of similar character to work set out in the Contract Bills but is not executed under similar conditions thereto and/or significantly changes its quantity, the rates and prices for the work so set out shall be the basis for determining the valuation and the variation shall include a fair allowance for such difference in conditions and/or quantity; valuation;

3. Where the additional or substituted work is not of a similar character to work set out in the Contract Bills, the work shall be valued at fair rates and prices.

If the valuation relates to the execution of additional or substituted work, which cannot be properly measured, then the work should be valued on a daywork basis in accordance with the 'Definition of Prime Cost of Daywork executed under a Building Contract' issued by the RICS and the Construction Confederation as current at the base date together with percentage additions to the prime cost at the rates set out by the contractor in the contract bills (clause 5.7).

Apart from the requirement under clause 52(1) in the ICE *Contract*, 7th edition, where the contractor is required to submit a quotation *without* reference to the contract rates or prices, the provisions for valuing variations under both the ICE *Contract*, 7th edition, and the JCT 2005 are very similar.

The most comprehensive review of the legal principles involved in valuing variations under the JCT 98 contract was undertaken by Mike Rycroft, director of James R. Knowles and Dr Issaka Ndekugri, director of the MSc Construction Law award at the University of Wolverhampton (Rycroft and Ndekugri, 2002). The article, which examined the contract provisions with an aim of providing guidance on practical implications and how to avoid or deal with essential pitfalls, has since been referred to by learned judges when dealing with legal disputes involving variations.

14.4   Contractual requirements – *The NEC Engineering and Construction Contract*, 3rd edition

The 3rd edition of the *NEC Engineering and Construction Contract* (NEC3) was published in June 2005.

One of the more radical changes introduced by NEC3 is the concept of 'Compensation Events', a term to denote any incidence of risk for which the client accepts liability under the contract. Clause 60.1 sets out 19 compensation events of which clause 60.1(1) on changing the works information is equivalent to the variations clauses under other standard forms of contract.

After discussing with the contractor different ways of dealing with the compensation event, which are practicable, the project manager may instruct the contractor to submit alternative quotations (clause 62.1).

Clause 62(2) requires the contractor to submit a quotation for the variation reflecting both the time and cost implication.

Clause 63 identifies that the changes to the prices are assessed as the effect of the compensation event upon:

- the actual defined cost of the work already done;
- the forecast defined cost of the work not yet done;
- the resulting fee.

Clause 63.6 identifies that assessment of the effect of a compensation event includes risk allowances for cost and time for matters which have a significant chance of occurring and are at the contractor's risk under this contract.

Quotations should be based on the assessment of *actual cost*, the definition of which is given in the contract. For example, 'Option B: Priced contract with bill of quantities', *Actual cost is the cost of components in the Schedule of Cost Components*. This in effect links actual cost and assessment of the compensation event back to the schedules which relate to 'Part Two – Data provided by the Contractor', again making the parties' *bargain* form the basis for valuation. The idea with the schedule of cost components is that conceptually the contractor is in the same position for a compensation event as when he tenders for the work (Mitchell and Trebes, 2005).

If there are time and disruption implications then a revised programme must be submitted.

Two viewpoints from practitioners on the practical issues involved in using the NEC schedule of cost components for valuing variations are contained within the NEC *Users' Group Newsletters*, issues 20 and 22.

14.5 Fixing the rate

Max Abrahamson, in his book *Engineering Law and the I.C.E. Contracts* (1979) (based on the ICE 5th edition), states that rate fixing 'is normally a give-and-take operation between the engineer and contractor'. Later he clarifies this: 'The basic consideration is that the contractor has agreed to do all work within the contract – original and varied – on the basis of his bill rates.'

Max Abrahamson recommends that when fixing the rates the parties should attempt to follow the following rules:

- *General principle*: try to follow the same principles that the contractor used in calculating his rates for the tender;
- *Fair valuation*: fair to both parties, that is, cost plus a reasonable percentage for profit, with a deduction for any proven inefficiencies by the contractor;
- *Market rate*: may be taken in consideration or used completely;
- However only in exceptional cases should the basis of valuation from the BofQ rates be abandoned.

The logic in using bill rates and prices for valuations is that the contract itself is founded on these rates, and since the contract contemplates variations, it is fair to both parties that bill rates should be used in valuations.

The contract gives very little practical guidance on the method of rate fixing. It will normally be necessary to break down the quoted rates into the various elements of plant, materials, labour and overheads in order to make appropriate adjustments.

Table 14.1 shows how the original BofQ rate can be used to establish a new rate after the issue of a variation order changing the description of the formwork from exceeding 1.22 m to a formwork 0.4–1.22 m wide. The new rate has been built up using the same rate for *materials*, with 10% addition on the *labour* and *plant* elements to reflect formwork to smaller areas. The same percentages for *site overheads* and *head office overheads* and *profit* as included in the original BofQ rate, 15% and 10% respectively, has been incorporated into the build up of the new *star rate*.

In order to adopt this approach, it is recommended that the employer's contract administrator/ QS obtain a breakdown from the contractor of the six most significant rates prior to the signing of the contract. This information is rarely requested in the UK at the pre-tender stage. However, under the Singapore post-1980 SIA contract it is a contractual requirement that contractors provide a make-up of prices (reported in Hudson, 1995, p. 946).

The author recollects visiting a contractor's office while working for the HKMTRC in Hong Kong in order to establish the basis of the tender build-up; there was one small problem – the

Table 14.1 Example of using BofQ rate to establish new rate ('star rate').

	BofQ item Formwork rough finish plain vertical exceeding 1.22 m (CESMM3 Item G145) (£)	New 'star rate' item Formwork rough finish plain vertical 0.4–1.22 m (CESMM3 Item G144) (£)
Labour	30.00	33.00
Materials	9.00	9.00
Plant	4.00	4.40
	43.00	46.40
Site overheads @15%	6.45	6.96
	49.45	53.36
Head office overheads @ 10%	4.95	5.34
	54.40	58.70

contractor was French and not surprisingly the estimate was in French. Thoughts of visiting the Japanese contractors for a similar tender breakdown were soon abandoned!

It can be said that the position for valuing variations under sub-clauses 52(3) and (4) under the ICE 7th edition is in line with most other standard forms, that is, the basis of the valuation is on the concept of price rather than cost. The rationale is that the rates identified in the contract bills or schedules of rates will form the basis, either directly or indirectly, for the value of the additional works.

The intention of the contractual provisions will normally be to maintain the competitive element in the valuation of variations as represented by the contract bills or schedules of rates – if used – so that the parties' *bargain* forms the basis of valuation. This means adopting the existing rates as the starting point for valuing variations, irrespective of whether they appear too high, too low or unreasonable for some other reason. This philosophy is confirmed in *Dudley Corporation v Parsons & Morrin (1959)* and *Henry Boot v Alstom (2000)* (see summaries at end of chapter).

Furst and Ramsey (2001) in Keating point out that *similar conditions* are those conditions which are to be derived from the express provision of the contract documents. Extrinsic evidence of, for instance, the parties' subjective expectations is not admissible.

However, in practice, contractors often argue that the work is not similar to the tendered work and fair rates should apply, often claiming that the work should be valued on a daywork basis and valued based on the actual records of resources used. Judge Bowsher QC in his judgement in the case of *Laserbore Ltd v Morrison Biggs Wall Ltd (1993)* CILL 896 had to decide the meaning of *fair and reasonable payments for all works executed*. He considered that the costs-plus basis was wrong in principle even though in some instances it may produce the right result. The appropriate approach was to adopt general market rates. The judge saw no objection in the use of FCEC (Federation of Civil Engineering Contractors) schedules of dayworks, provided there was no duplication in payment for insurances and head office charges.

In the *Tinghamgrange* case (discussed under the heading 14.7, Some other relevant legal cases) the Court of Appeal held that on a contract based on the ICE 5th edition a fair valuation under clause 52(1) included compensation to the main contractor for a loss of profit payment to a subcontractor in respect of the cancellation of an order resulting from an engineer's variation.

In the *Weldon Plant* case (discussed under the heading 14.7, Some other relevant legal cases) it was held that a fair valuation under clause 52 of the ICE 6th edition should be based upon the reasonable cost of carrying out the work if reasonably and properly incurred. His Honour

Judge Humphrey Lloyd QC considered that a fair valuation must include something on account of each of the elements which is ordinarily to be found in a contract rate or price; elements for the cost of labour, the cost of plant, the cost of materials and the cost of overheads and profit, while time-related overheads (preliminaries) might require to be proved.

EXAMPLE

A contractor has inserted a rate of £28 per m^2 for formwork to a retaining wall. The quantity was 576 m^2 based on 24 bays 8 m long by 3 m high.

The engineer issued a variation order reducing the retaining wall to 16 bays, i.e. 384 m^2 after the shutter had been ordered but before the work commenced on construction of the wall.

Calculate a revised rate for this varied work making necessary assumptions (8% site overheads/10% head office overheads, profit and risk). The contractor has informed the engineer that in pricing the item he allowed for a purpose-built steel shutter 8 m x 3 m at a capital cost of £6,000.

Solution

First, it is necessary to analyse the original rate of £28.00 per m^2. Unlike a traditional build up of a rate we will need to start at the end and work backwards.

	£
BofQ rate (based on 576 m^2)	28.00
Less 10% head office overheads and profit	
(divide by 110 then multiply by 100)	2.55
	25.45
Less 8% site overheads	
(divide by 108 then multiply by 100)	1.89
Net rate	23.56
Material: purpose-built shutter	
£6,000 divided by 576 m^2	10.42
Labour and plant:	
Erect and strike including all necessary cranage	13.14*

Now that we have analysed the original rate, we can build up the rate for the varied work using the same approach as above.

Labour and plant:	
Erect and strike including all necessary cranage	13.14* (as original)
Material: purpose-built shutter	
Contractor is entitled to recover full cost	
£6,000 divided by 384 m^2	15.63
	28.77
Add 8% site overheads	2.30
	31.07
Add 10% head office overheads and profit	3.11
New rate for formwork (based on 384 m^2)	34.18/ m^2

The rate should be confirmed on a *star rate* form and endorsed by authorized representatives from both the employer and contractor. This logical approach complies with the contract terms and should satisfy the auditor if the project is subject to audit.

Practical considerations

Variations have the potential to generate significant additional costs, delay and disruption to even the best-planned construction project. The following 11 points are some of the key factors which should be considered for inclusion in a quotation submitted before the work is executed or in assessing a *fair* valuation after the event:

1. *General items*: including revised method statement, effect on the critical path, revised production rates, out-of-sequence working, restricted access, summer-to-winter working, changed nature of ground, temporary works, additional cleaning, late payment and financing etc.;
2. *Labour*: including uneconomic working, difficulty of access, attraction money, additional bonus, overtime payments, shift work, accommodation and welfare etc.;
3. *Construction equipment*: including additional mobilization/demobilization, transport costs, additional scaffolding/hoisting/cranage, standing charges, additional payments to operators, working out of sequence etc.;
4. *Materials*: including additional costs of late orders, additional procurement costs, airfreight, premium costs, small quantities, excessive waste, potential breakages and additional testing etc.;
5. *Subcontractor costs*: including additional costs in expediting, possible additional visits to the subcontractors' factory/yard/works in the UK or overseas;
6. Inflation effect if work executed at a later date;
7. Additional costs in design work and reprogramming;
8. Cost of preparing the quotation;
9. Additional time-based preliminaries/site overheads;
10. Contingencies for contractor's risk;
11. Head office overheads and profit.

This above list is significant as it indicates the extent of the potential differences in the positions of the two main parties. A typical scenario is when the contractor wants to include many of the above items in the quotation while the client's QS is unable to agree and the variation ends up being valued using the traditional approach. The parties need to keep talking and negotiating in an open manner within a true spirit of trust and cooperation.

Feedback from research

Dr Monty Sutrisna investigated the quotation mechanism for pre-pricing variations on civil engineering works (Sutrisna *et al.*, 2004). Feedback was obtained from 95 participants with a wide range of experience on the administration of civil engineering projects, however exposure to ICE 7th was acknowledged by many as limited with the emphasis in the sector shifting to use of customized versions of the NEC ECC 2nd edition. The respondents identified that the quotation mechanism has been perceived effective only in certain conditions. The perceived problems were two-fold: limited time available to produce the quotation and calculation of appropriate risk allocation.

The following best practice in connection with quotations for variations which had the greatest chance of being accepted was also identified:

- Basis of calculation of overheads in quotation – use original BofQ overheads;
- Basis of pricing level to be applied – pricing level at the time of preparing BofQ;
- Basis of profit level – original BofQ profit level;
- Contingency for risk – based on risk-analysis assessment.

Dr Denise Bower's research (2000) was developed based on a technique known as *Impact* used by Fluor Daniel Ltd in the process-plant sector and was tested on low-risk, low-technology civil engineering projects. The technique enables the parties to calculate the *influence curve* for a project enabling the calculation and prior agreement of the indirect charges of variations based on the type of work involved. This technique would seem to have considerable merit particularly on partnering projects where there is a long-term relationship between the parties.

SG Revay's research in 1992 based on an analysis of 175 projects in Canada found that

All in all the real cost of variations resulting from the incorrect or incomplete bid documents represented an average of 33.5% of the original contract price. Simply stated, the real cost of the variations injected during the currency of the project was triple that of the direct.

14.6   *Quantum meruit* claims

Quantum meruit, Latin for *as much as he has deserved*, can arise in two forms: contractual and quasi-contractual. Quasi-contractual actions embrace a range of claims based upon unjust enrichment; therefore, they reside within the law of restitution.

Contractual *quantum meruit* claims may arise in the following situations:

- Where the contract provides for the payment of a reasonable sum;
- Where the contract does not stipulate the price to be paid or the contractual pricing mechanism fails;
- Where part performance has been accepted;
- Where an innocent party elects to treat the contract as discharged following a repudiatory breach by the other party.

The circumstances in which restitutionary *quantum meruit* are likely to arise are as follows:

- Work is carried out under the erroneous assumption that a contract exists.
- Work is carried out in anticipation of a contract being concluded.
- Work is performed outside the scope of the contract.

Work is carried out under the erroneous assumption that a contract exists

An example is the case of *Peter Lind & Co Ltd v Mersey Docks and Harbour Board (1972)* 2 Lloyd's Rep. 234. The civil engineering contractor Lind had submitted two tenders to the Harbour Board. One of the tenders was for a fixed price and the other incorporated a fluctuation clause to account changes in the price of labour and materials. The Board accepted Lind's tender without specifying which one. Nonetheless, Lind carried out the works and claimed payment on a *quantum meruit* basis.

The court held that there was no concluded contract since it was not clear which tender had been accepted and, therefore, Lind was entitled to be paid on a *quantum meruit* basis.

Work is carried out in anticipation of a contract being concluded

It is fairly common for work to begin while negotiations are continuing over critical issues such as price, scope of works and date of completion. If these matters remain unresolved, a contract is likely to arise even though work is underway.

The best illustration is *British Steel Corporation v Cleveland Bridge & Engineering Co Ltd (1984)* 1 All ER 504. Cleveland Bridge was involved in the construction of a bank in Saudi Arabia. For this purpose it required the manufacture and supply of steel nodes. Accordingly it issued a letter of intent to British Steel for the supply and delivery of the nodes. The letter requested British Steel to 'proceed immediately with the works pending the preparation and issuing to you of the official form of sub-contract.'

In the event, there was neither an agreement on price or delivery dates and nor did British Steel receive the *official form of sub-contract*. It sued Cleveland Bridge for £229,838; the amount being the price of the nodes. Cleveland Bridge, while admitting liability in the sum of £200,853, issued a counterclaim for £867,736 on the basis that, in breach of contract, British Steel had delivered the nodes late and out of sequence. The alleged contract comprised the letter of intent, a subsequent telex from Cleveland Bridge dealing with delivery sequence and, finally, British Steel's conduct in delivering the nodes. British Steel claimed that no contract had come into existence and therefore the counterclaim was misplaced.

Judge Robert Goff held that British Steel was entitled to its claim on a *quantum meruit* basis. The contract between the parties was still in a state of negotiation; there was no agreement on critical issues especially the terms of the proposed subcontract.

Work is performed outside the scope of the contract

When work is carried out outside the scope of the contract, remuneration for such work can be obtained only on a *quantum meruit* basis. The case of *Costain Civil Engineering and Tarmac Construction Ltd v Zanen Dredging and Contracting Co Ltd (1997)* CILL 1220 is instructive.

The Welsh Office appointed Costain/Tarmac Joint Venture (JV) as main contractors under an ICE 5th edition contract for the construction of the A55 Conwy bypass and river crossing in North Wales. Part of the works involved dredging a trench in the bed of the estuary of the River Conwy, into which six prefabricated tunnel elements, made of reinforced concrete, were to be immersed and jointed together to form the carriageway of the bypass. These tunnel elements were constructed on-site in what was known as a casting basin, and when completed the casting basin was flooded and the tunnel elements floated out to their positions in the estuary where they were sunk into place. Zanen were engaged as dredging subcontractors under the Blue Form subcontract.

The contract provided options for dealing with the casting basin once works were complete. One proposal was that the basin should be backfilled, and this had been priced by the Costain/Tarmac JV as a saving in excess of £1 million, on the basis that it would be cheaper for them to backfill the casting basin rather than to remove the spoil and dispose of it off-site.

As an alternative, Crown Estates, who were not a party to the main contract, wanted to build a marina using the flooded casting basin. Accordingly they entered into agreements with the Welsh Office and with Costain/Tarmac to the effect that the contractor would credit the Welsh Office with the £1 million previously mentioned, but approximately £2.5 million would be paid by Crown Estates for additional works around the perimeter of the marina.

The JV instructed Zanen to carry out additional works, which the Zanen considered outside their subcontract, but the JV stated they were not and continued giving instructions relating to the marina works. The court agreed with Zanen that the work to the marina was outside the terms of the original subcontract and therefore fell to be evaluated using a *quantum meruit* approach.

The JV suggested that the correct approach was to reimburse the cost to the subcontractor of executing the marina works (£380,000) and allow a reasonable uplift in respect of its overheads and profits of 10%. His Honour Judge Wilcox did not accept this position and considered that as Zanen had executed the work, which had been wrongly instructed under protest, the assessment should be based on the principles of restitution and unjust enrichment. He awarded Zanen reimbursement based on *quantum meruit* calculated by reference to the cost (£380,000) together with a portion of the substantial profit made by the JV for the marina works (a further £380,000) which he considered reflected the benefit to the JV of having those works executed by a subcontractor whose resources were already mobilized on-site.

14.7 Some other relevant legal cases (reported in date order)

Simplex Concrete Piles v St. Pancras Borough Council (1958) *14 BLR 80*

Under an RIBA contract, the contractor undertook to carry out design and construction of piling for the foundations on a block of flats and guaranteed to satisfy certain tests. In the event, conditions made it impossible to satisfy the tests and the contractor suggested two alternative methods of piling – one method to be carried out by themselves and another involving a subcontractor.

The contractor submitted the two prices to the architect and received the following response 'we are prepared to accept your proposals that the piles . . . should be of the bored type in accordance with quotations submitted by the (the sub-contractor).'

It was held that the architect's letter was an architect's instruction involving a variation in design and the contractor was entitled to be reimbursed based on the subcontractor's quotation and not as the original tender price.

Dudley Corporation v Parsons and Morrin Ltd. (1959) *[Court of Appeal]*, Building Industry News, *17 February 1967*

In this case an extra-only item for excavating 750 cubic yards in rock was priced at £75 total, that is, 2 shillings (10 pence) per cubic yard. A fair and reasonable price would have been £2 per cubic yard.

It was not known beforehand whether rock would be met, but, in fact 2,230 cubic yards of rock was excavated.

The architect (under a 1939 RIBA with Quantities Form) valued the excavation at 750 cubic yards at 2 shillings (i.e. the original extension of £75) and the balance at £2 per cubic yard. It was held that this approach was incorrect and that the contractor was entitled to 2 shillings per cubic yard only for the whole of the quantity excavated.

A.E. Farr v Ministry of Transport (1965) *3 E.R. 88; 5 BLR 94*

This case concerns a roadworks project executed under the ICE *Conditions of Contract* 4th edition with a BofQ prepared in accordance with a 'standard method of measurement'.

Clause 12 in the conditions stated that the rates and prices in the BofQ were to cover all the contractor's obligations under the contract.

The BofQ specified that measurement of excavation be based on the net volume with any additional excavation required for working space paid for as a separate item based on the sum areas of the excavation.

In the event no separate item for working space was measured, though two specific items for working space were included in a part of the BofQ relating to subsidiary parts of the works. The contractor claimed that he should be paid for working space whenever it was reasonable to excavate outside the net plan area.

It was held by a majority in the House of Lords (overruling the decision of the Court of Appeal) that the quoted words amounted to a promise to pay the contractor extra for all working space required, whether or not described in a special item in the BofQ.

Note that this is in contrast with the modern provision of the *Civil Engineering Standard Method of Measurement* 3rd edition where working space is not measured separately as it is 'deemed included' in the contractor's excavation rates (CESMM3 Class E Coverage Rule C1).

Mitsui Construction Co v Att. Gen of Hong Kong (1986) *33 BLR 1*

The conditions of contract were specifically produced by the Hong Kong Government and were based on the ICE 4th and JCT63. The two-year project involved the construction of a tunnel 3227 m long and 3.6 m diameter. The ground conditions were extremely variable and the engineer specified five different types of tunnel lining suitable for the different ground conditions.

In the event, the ground conditions were far worse than expected and the contractor was required to construct far more of the heavily designed tunnel section (2448 m compared to 275 m billed) and much less of the unlined section. An extension of time (EoT) of two years was granted.

The engineer argued that the changes in quantity were not the result of an official variation order and the contractor was paid at the rates in the BofQ. The contractor claimed that the increased quantities amounted to variations and that revised rates should apply. The Hong Kong government took the view that the engineer had no power to revise the rates.

The Privy Council took 'a sensible and business-like approach' and found in favour of the contractor, stating that the engineer did indeed have the power to fix a revised rate.

English Industrial Estates v Kier Construction Ltd (1991) *56 BLR 93*

Two contracts were let to Kier in June and November 1987 for the reclamation at the former Dunlop factory at Speke in Liverpool. The excavation for both subcontracts was sublet to J&B Excavation Ltd.

In the specification the contractor was given a choice for structural fill of either using material arising from demolition or importation.

The contract required the contractor to submit his programme and method statement with his tender. The contractor's method statement showed that the excavation subcontractor intended to crush only *suitable material arising from the demolition* with the remainder removed from site.

In January 1988, the engineer wrote to the contractor instructing him to crush *all* hard material arising from site. The subcontractor claimed compensation for the losses due to the additional costs involved.

The arbitrator held that the contractor's method statement was a contract document and the engineer's instruction was thus a variation under clause 51 of the ICE 5th.

On appeal, the High Court agreed with the arbitrator's decision.

Tinghamgrange Limited v Dew Group Limited (1995) *47 Con LR 105*

This case involved North West Water who commissioned works at its Oswestry Water Treatment Works. Part of the works required the removal and replacement with new, precast concrete under-drainage blocks.

The precast concrete blocks had to be specifically manufactured early in 1989 for inspection by North West Water, prior to the contractor Dew placing an order.

The order was placed in April 1989 for 282,354 precast concrete drainage blocks at 90 pence each. Dew's order contained a condition allowing cancellation and restricting the supplier Tinghamgrange's right in respect of any claim.

Ultimately, North West Water instructed Dew to cancel the order, as a change of specification was required. Tinghamgrange claimed loss of profit on the blocks ordered but not delivered as a result of the cancellation. The claim was passed onto North West Water as part of the costs associated the variation.

North West Water accepted the cost of the special mould manufactured for the purpose of fulfilling Dew's order but rejected the claim in respect of loss of profits.

There was no dispute that the change was a variation and it was accepted that the substitution of tiles for concrete blocks involved work, which was not of a similar character. This required the engineer to value the new works using BofQ rates as the basis so far as reasonable and if they could not provide a proper basis for valuation then a fair valuation was required.

By a majority of 2:1, the Court of Appeal allowed Dew to recover the loss of profit; the majority view was that Tinghamgrange's loss was an integral part of Dew's costs and to exclude that element from a valuation of the work would be unfair.

Henry Boot Construction Ltd. v Alstom Combined Cycles Ltd. (2000) *TLR April 11*

In this case, Alstom employed Henry Boot to carry out some civil engineering works at a new power station at Connah's Quay in Wales. The power station comprised four combined cycle turbines; each turbine comprised a turbine hall, a heat-recovery steam generator and a cooling tower. The ICE *Conditions of Contract* 6th edition applied.

During pre-contract negotiations, Boot submitted a price of £250,880 for temporary sheet piling to trench excavation in the turbine-hall area, and this price was incorporated into the contract.

During the course of the works the engineer issued variation orders instructing temporary sheet piling to trench excavation in the heat-recovery steam generator area and the cooling-tower area.

In the event, it was identified that Boot's price of £250,880 had been calculated in error including sheet piling to both the turbine hall and the heat-recovery steam generators, although the contract was clearly entered into on the basis that it was for the turbine hall alone.

The decision primarily concerns the application of clause 52(1)(b) in the ICE *Conditions*, dealing with the valuation of variations; the crucial part of the clause is set out below:

> Where work is not of similar character or is not executed under similar conditions or is ordered during the Defects Correction period the rates and prices in the Bill of Quantities shall be used as the basis for valuation so far as may be reasonable failing which a fair valuation shall be made.

Boot argued that the additional work should be valued using contract rates. Alstom considered that this approach was not reasonable and would result in an unjustified windfall for Boot; Alstom argued that a fair valuation should be made. Initially the matter went to arbitration with the arbitrator agreeing with Alstom. However on appeal the case was heard in the Technology and Construction Court before His Honour Judge Humphrey Lloyd QC, a most experienced construction lawyer.

The judge stressed the importance of the contract rates, and the fact that they cannot be avoided simply because one party is dissatisfied with them. 'The contract rates were sacrosanct, immutable and not subject to correction. The fact that a rate or price, which would otherwise be applicable, may be considered too high or too low is completely immaterial.'

The judge quoted with approval Max Abrahamson in his *Engineering Law* and the ICE *Contract* (p. 185): 'It is not unreasonable to apply rates as a basis for applying varied work merely because the rates are mistaken . . . . What is reasonable is to be decided by reference to the nature of the original and varied work, not extraneous conditions.'

The judge held that it was not a windfall for Boot – it was 'all part of the risks of contracting which produced thrills as well as spills'.

The Court of Appeal dismissed Alstom's appeal and confirmed the decision of Judge Humphrey Lloyd QC. The Court of Appeal also confirmed that clause 52(2) (which permits the engineer to vary a rate if the nature or amount of the variation is such as to make the rate inappropriate) did not justify displacing the rates themselves because they were inserted by mistake or were too high or too low or otherwise unreasonable.

The decision confirms the well-established view that a contractor will be held to his rates and prices in the contract for both original and varied work. It also illustrates the serious consequences of failure to pick up errors at tender stage.

Postscript: In a later case of *Aldi Stores Ltd v Galliford (UK) Ltd (2000)* the principle of using BofQ rates in valuing variations was confirmed. However in this case it had the opposite effect and the use of BofQ rates resulted in a significant loss to the contractor.

Weldon Plant Ltd v The Commission for New Towns (2000) *TCC BLR 496*

Weldon Plant Ltd entered into a contract, based on the ICE *Conditions*, 6th edition, for the construction of Duston Mill Reservoir. Material to be excavated consisted of clay and gravel. Under the terms of the contract, Weldon could excavate below the designed level of the reservoir bed to obtain additional gravel, which it was entitled to sell – the contractor priced this work at a negative rate.

However the engineer issued an instruction, which required Weldon to excavate all the gravel below the bed and to backfill with clay to the design level. The engineer valued the instruction as a variation and wished to apply the negative rate also awarding an extension of time of over seven weeks.

Weldon Plant disputed the engineer's findings and referred the matter to arbitration and later to the Technology and Construction Court.

The arbitrator found that the BofQ rates could not be applied in this instance and that a fair rate should be used. However, having decided upon this, the arbitrator refused to include in the fair rate any provision for profit or contribution which the varied work might have made to the fixed and running overheads of the business. His logic was that the contractor could not recover these unless he could prove that he had lost an opportunity to earn extra profit or overheads elsewhere by reason of the variation. As no such proof had been provided, fair rates and prices should exclude these elements.

The Technology and Construction Court disagreed with the arbitrator. In the court, the judge took the view that in evaluating a fair valuation the calculation should be based on the reasonable costs of carrying out the work if reasonably and properly incurred.

Attention was drawn to what *Keating on Building Contracts* has to say on the constituent elements of a fair valuation. It states that useful evidence may include a calculation based on the net cost of labour and materials used, plus a sum for overheads and profit. The judge considered that a fair valuation should include these elements and referred the case back to the arbitrator for him to include in his valuation an amount for overheads and profit.

The judge also expressed approval, in passing, of the principle that fair rates and prices should, if need be, be tempered so as not to fall too far out of line with the BofQ rates.

14.8   Conclusion

Valuing variations based on *cost* as opposed to *price* at first sight seems a radical shift in philosophy. However, in practice, the only real effect of the changes in the valuation of variations provision from the pre-2000 standard forms of contract to the ICE 7th edition, JCT 2005 and the NEC3 is that there is an intention to agree the valuation of variations in advance where possible.

In principle, the contractor could forward a quotation irrespective of *actual* likely cost and disruption. In reality such a quotation is likely to be rejected for being unreasonable and not 'in the spirit' of the contract conditions; indeed Dr Monty Sutrisna's research confirms that quotations are only effective in certain circumstances. In this instance, the effect of the rejection of the quotation is to bring into play provisions for valuation, which have existed for some time thus returning to the traditional variation rules.

Despite the foregoing it is appropriate to highlight the fundamental change in philosophy adopted by the new contracts. The NEC3 has clearly adopted a partnering-type approach and this has been followed to some extent initially by the ICE 7th and latterly by the JCT 2005. This should have a bearing when the contractor submits the quotation and the project manager/engineer/architect/contract administrator assesses it; indeed experience to date would suggest this is the case.

14.9   Questions

1. Identify the different opinions offered by experts on the definitions of *similar character/ conditions* and *fair rates, prices and allowances* as reported in Rycroft and Ndekugri (2002). Compare and contrast how these opinions might be applied in the case of changes to a deep basement for a pumping station on a greenfield site executed under either the ICE 7th, the JCT 2005 SBC/Q or the NEC3 *Contract*.
2. Identify the possible difficulties and solutions in valuing variations using the 'schedule of cost components' under the NEC3 *Contract*.
    Source: Paul Pavia, Franklin and Andrews, *NEC Users' Group Newsletter*, issue no. 20; Bryan Tyrell, Currie and Brown, *NEC Users' Group Newsletter*, Issue No. 22, available online at: www.newengineeringcontract.com
3. Identify those factors which should be considered by the client's project manager in developing an effective strategy to reduce/minimize the impact of variations on a construction project.
    You may make any assumptions concerning the type and nature of the project.
4. Where work is omitted from the contract by way of a variation can a contractor/ subcontractor claim for loss of profit?
5. The case of *Costain and Tarmac JV v Zanen Dredging* (1997) 85 BLR 77 concerned the construction of the Conwy Bypass and the river crossing in Wales. Zanen argued that work to a marina was outside their contract and should be valued on a *quantum meruit* basis. What is meant by *quantum meruit* and what are the guidelines for valuing work on this basis?
6. Critically review and summarize in 250 words the article by Denise Bower 'A systematic approach to the evaluation of indirect costs of variations' in *Construction Management and Economics* (2000), vol. 18, 263–268.

Bibliography

Abrahamson, M.W. (1979) *Engineering Law and the I.C.E. Conditions of Contract*, 4th edition, Elsevier Applied Science Publishers

Bower, D. (2000) 'A systematic approach to the evaluation of indirect costs of contract variations', *Construction Management and Economics*, vol. 18, pp. 263–268.

Broome, J.C. (1997) 'Best practice with the new engineering contract', *Proceedings of Institution of Civil Engineers,* 120, May, pp. 74–81

Eggleston, B. (2001) *The ICE Conditions of Contract*, 7th edition, Blackwell Scientific Publications.

Furst, S. and Ramsey, V. (2001) *Keating on Building Contracts*, 7th edition, Sweet & Maxwell

Gray, C., Hughes, W. and Bennett, J. (1994) *The Successful Management of Design, A Handbook of Building Design Management*, The University of Reading

Haswell, C.K. (1963) 'Rate fixing in civil engineering contracts', *Proceedings of Institution of Civil Engineers*, vol. 24, February, pp. 223–234

Michell, B. (2003) *Jubilee Line Extension: From Concept to Completion*, Thomas Telford

Mitchell, B. and Trebes, B. (2005) *NEC Managing Reality: Book 4 Managing Change*, Thomas Telford

Revay, S.G. (1992) 'Can construction claims be avoided?' in Fenn, P. and Gameson, R. (eds) *Construction Conflict and Resolution*, International Construction Management Conference, Manchester, E&FN Spon

Rycroft, M. and Ndekugri, I. (2002) 'Variations under the JCT Standard Form of building contract', *Construction Law Journal*, vol. 18, no. 4, pp. 310–333

Sutrisna, M., Potts, K. and Proverbs, D. (2004) 'Quotation mechanism for pre-pricing variations in civil engineering projects: A quest for best practice', *Journal of Financial Management of Property and Construction*, vol. 9, no. 1, pp. 13–25

Useful websites

www.atkinson-law.com
www.brewerconsulting.co.uk
www.jrknowles.com
www.newengineeringcontract.com

15   Claims management

15.1   Introduction

Settling delay claims is a challenge to both clients and their advisors and contractors as claims management requires knowledge of construction technology, construction law (including relevant case law), the conditions of contract, contract administration, project-planning systems and the psychology of negotiation.

The architect/contract administrator or project manager/engineer does not have an easy task when settling extension of time claims. His Honour Judge Richard Seymour QC in the legal case of *The Royal Brompton Hospital NHS v Watkins Gray International (UK) (2000)* identified that the architect needs three basic skills.

The first skill is *construction knowledge*. Judge Seymour observed that the construction of a modern building involves the carrying out of a series of operations, some of which can be undertaken at the same time as others, but many can only be carried out in sequence. It is not therefore immediately obvious which operations have an impact upon others and which delays affect the ultimate completion date. The architect must therefore have an in-depth knowledge of construction and the interrelation between trades and construction operations.

The second skill is *an understanding of programming techniques*. Judge Seymour observed that in order to make an assessment of whether a particular event has affected the ultimate completion of the work, rather than just a particular operation, it is necessary to consider what operations, at the time when the event happens, are critical to the forward progress of the work as a whole. The architect will usually have to adopt an appropriate programming technique to analyse the effect of various events. There are a number of established methods of analysis, but each is likely to produce different results to others, sometimes dramatically different results.

Most importantly, the accuracy of any of the methods in common use depends upon the quality of the information used. It is much more difficult to establish the critical path if one does not know how the contractor planned the job. Not only that, but the critical path may well change during the course of the works, and almost certainly will do if the progress of the works is affected by some unforeseen event.

The third skill is *contractual awareness* – the architect must understand the relevant contractual provisions and be up-to-date on decided cases.

To add to the architect's difficulties, as observed by Judge Seymour, often the contractor gives a written notification of delay regardless of whether it really thought that the event would cause delay to the completion of the works. Notices are given every time anything alters or anything happens which could conceivably delay any individual activity. Although from a contractor's point of view adopting such a practice has the advantage that he is covered, no matter how things should turn out, it does make life difficult for the architect. In addition, the contractor often makes exceptionally optimistic predictions of the extent of the likely delay caused by the matters that it notified.

Claims are inevitable on most large-scale construction projects and are usually motivated by a single cause – the contractor, or subcontractor, anticipates spending, or actually spends, more money than they expected and they believe someone else is responsible.

It is important to identify some of the main reasons leading to the submission of claims by contractors these might include the following:

- inadequate time and planning before the project commenced on site;
- inviting tenders on incomplete drawings;
- introducing extensive revisions throughout the project;
- inadequate site investigation – particularly on civil engineering works involving deep basements, piling, earthworks, tunnelling etc. In the author's experience a claim for unforeseen ground conditions is encountered on most large-scale projects;
- extensive changes to standard forms shifting the risk to the contractor often lead to claims – standard forms of contract are tightly integrated documents;
- client's interference with the timing and sequence of construction.

15.2   Terms in contract conditions

The terms in the relevant contract are obviously critical; these will identify the grounds for extensions of time and cost recovery and state the procedures and relevant timescales for notifications and submissions. There is likely to be a close link between making a claim for delay and recovering the additional costs incurred by the contractor. Delay claims are therefore of considerable commercial importance to contractors.

The effect of extending time is to maintain the contractor's obligation to complete within a defined limit and failure by the contractor to do so leaves him liable for damages, either liquidated or general according to the terms of the contract.

Standard JCT form of Building Contract, 2005 Edition (JCT 05)

An architect/contract administrator, who is appointed by the employer and acts as his agent, administers JCT 05; however it is noted that the architect is required to act fairly and reasonably towards the contractor. Extensions of time are dealt with under clause 2.27 with clause 2.29 listing 13 categories of delay, called *relevant events* that give rise to an extension of time in the event that they occur. The key issues identified are as follows:

- When it becomes apparent that the progress of the works is being or is likely to be delayed then the contractor should forthwith notify the architect/contract administrator of the cause of delay and identify whether it is a relevant event.
- The contractor is required to provide with the notice, or as soon as possible after the notice, particulars of the event and an estimate, if any, of the expected delay to the completion of the works or any section beyond the relevant completion date.
- Upon receipt of the notice and any further particulars, the architect/contract administrator is required to decide whether in his opinion any of the events notified are relevant events and whether as a result of such events the works are likely to be delayed beyond the completion date. If they so decide, they are then required to give a fair and reasonable extension of time to the contractor.

In practice, it is quite often the case that the architect will wait for full particulars of the actual effect of the event before making a decision, by which time the event and its consequences may be long past and the actual effect may be measured with greater certainty. It is quite common for architects to wait until after completion before making such grants of extension of time, but this is clearly not what the clause intended (Burr and Palles-Clark, 2005).

The claim for loss and expense is dealt with separately under clause 4.23 which lists only five *relevant matters*.

ICE Conditions of Contract, Measurement Version, *7th edition, 1999 (ICE 7th)*

The ICE 7th is administered by the engineer who is appointed by the employer and acts as his/her agent, but is also required to act fairly towards the contractor and act as an independent quasi-arbitrator in the events of disputes: 'Should the Engineer consider that the delay suffered fairly entitles the Contractor to an extension of time for the substantial completion of the Works . . . '.

Items giving grounds for an extension of time under the ICE 7th include the following:

1. any variation ordered under clause 51(1);
2. increased quantities referred to in clause 51(4);
3. any cause of delay referred to in the conditions, for example, clause 7(4) late drawings;
4. exceptionally adverse weather conditions;
5. any delay, impediment, prevention or default of the employer;
6. other special circumstances of any kind whatsoever.

An extension of time under the ICE 7th gives no entitlement to payment to the contractor and the question of whether a particular delay is reimbursable or non-reimbursable is properly determined from the cause of delay and proof of cost arising.

The following issues give potential grounds for extension of time and extra cost under the ICE 7th:

Clause	Subject
5	Documents mutually explanatory
7(4)	Further drawings – delay in issue
12(6)	Unforeseen conditions (including profit)
13(3)	Engineer's instructions
14(8)	Revised method or programme
31(2)	Facilities for other contractors
32	Fossils
40(1)	Suspension of the works
42(3)	Late possession of the site
59(4)	Nominated subcontractor's default.

15.3   Legal requirements of claims submission

Delay analysis is necessary for two main reasons:

1. to demonstrate entitlement to extension of time and hence relief from liquidated damages;
2. to demonstrate entitlement to the costs of prolongation.

It is relevant to consider the observations made by judges when dealing with the settlement of construction disputes.

In the later case of *John Barker v London Portman Hotel (1996)* the judge observed:

[the Architect] did not carry out a logical analysis in a methodical way of the impact which the relevant matters had or were likely to have on the Plaintiff's planned programme. He made an impressionistic, rather than a calculated assessment.

In the more recent case of *Balfour Beatty v Borough of Lambeth (2000)* His Honour Judge Humphrey Lloyd QC observed:

By now one would have thought that it was well understood that, on a contract of this kind, in order to attack (a non-completion certificate or an EoT) the foundation must be the original programme (if capable of justification and substantiation to show its validity and reliability as a contractual starting point) and its success will similarly depend on the soundness of its revisions on the occurrence of every event, so as to be able to provide a satisfactory and convincing demonstration of cause and effect. A valid critical path (or paths) has to be established both initially and at every later material point since it (or they) will almost certainly change.

Some means has also to be established for demonstrating the effect of concurrent or parallel delays or other matters for which the employer will not be responsible under the contract.

The contractor's claim must fill certain legal criteria:

- The claim must prove that a loss has been suffered.
- The claim must show that the loss arose as a result of the relevant acts or omissions.
- The contractor is under the legal burden of proving the link between the event, or the cause, and the delay to completion, or the effect; they must therefore be able to demonstrate the cause and effect.
- The legal quantum of the claim must be established. The measure of damages in common law remain as stated in *Robinson v Harman (1848)*:

The rule of common law is that where a party sustains a loss by reason of a breach of contract, he is as far as money can do it, to be placed in the same situation, with respect to damages, as if the contract had been performed.

This comment indicates that the contractor is not entitled to earn additional profit on the claim.

- It must be shown that the loss could not have been mitigated by reasonable conduct; for example, the contractor should remove mobile construction equipment if no work is being carried out.
- The losses must not be seen to be too remote. The principles concerning a common-law damages claim were set down in *Hadley v Baxendale (1854)*. In this case the court laid down two situations where the defendant should be liable for loss caused by a breach of contract:

    a.   Loss which would arise naturally, *according to the usual course of things*, from their breach;
    b.   Loss *as may reasonably be supposed to have been in the contemplation of the parties at the time when they made the contract, as the probable result of the breach of it.*

15.4   Contractor's programme

The contractor will normally be required to produce a construction programme at the commencement of the project and both parties will rely on this programme to justify any extensions of time. However, unless the programme is submitted with the bid it is unlikely to become a contract document and the client's representative will be under no obligation to accept it as a basis for payment.

The programme should preferably be in network format in order that the logic can be checked and the critical path established. Normally extensions of time would be awarded only for any items, which are delayed and are on the critical path. The programme should be linked with the method statement and record key dates for information required from the client. In order to avoid confusion, some enlightened clients specify the software to be used by contractors/subcontractors in order to ensure compatibility with their own systems.

The contractor may have included *float* within the programme to allow for any time for which they are responsible, for example, inclement weather. Traditionally this float has been considered the contractor's own and should not be utilized by the engineer without compensation to the contractor. However in the case of *Anson Contracting Limited v Alfred McAlpine Construction Isle of Man (1999)* Judge Hicks considered that any float should be considered on a first-come–first-served basis and that McAlpine, not having suffered any loss, was not allowed to recover from its subcontractors a hypothetical loss it would have suffered had the float not existed.

15.5   Concurrent delays

The courts generally favour the *common-sense* approach when dealing with matters of causation. However, in practice, there may be competing causes of delay. For example, delays caused by the client and entitling the contractor to additional time and cost (e.g. late instructions) may occur at the same time as a delay due to exceptionally bad weather (normally time only) or breakdown of the contractor's plant (contractor's risk).

Keating on Building Contracts, 7th edition (Ramsey *et al.*, 2001) offers a number of alternatives for settling these complex issues.

The Devlin approach

If a breach of contract is one of two causes of loss, and both causes co-operating and are of approximately equal efficacy, the breach is sufficient to carry judgement for the loss.

The Devlin approach if applied to delays would always come down in the contractor's favour if one of the competing causes of delay was a breach of contract on the part of the employer or the engineer or architect acting on his behalf.

The dominant-cause approach

If there are two causes, one the contractual responsibility of the defendant and the other the contractual responsibility of the plaintiff, the plaintiff succeeds if they establish that the cause for which the defendant is responsible is the effective, dominant cause.

Which cause is dominant is a question of fact, which is not solved by the mere point of order in time, but is to be decided by applying common-sense standards.

Keating on Building Contracts supports this approach:

The dominant-cause approach is supported as indicated above by great authority of weight in insurance cases. It is thought that the principles, so far as they apply, apply to contracts generally. It is accordingly submitted that the dominant-cause approach is or should be the correct approach, as the law now stands, for Case C and for Case B also, unless exceptionally the contract on its true construction provides explicit answer without sophisticated analysis.

Case B, as described in *Keating on Building Contracts*, concerns claims for payments under the contract for delay resulting from variation instructions where there is a competing cause of delay which could be no one's fault or the contractor's own delay in breach of contract. Case C provides for the same situation but where the contractor is instead claiming damages for breach of contract.

However it is noted in an earlier case *H Fairweather & Co v London Borough of Wandsworth (1987)* the court considered that the dominant-cause approach was not correct.

The burden-of-proof approach

If part of the damages is shown to be due to a breach of contract by the plaintiff, the claimant must show how much of the damage is caused otherwise than by his breach of contract, failing which he can recover nominal damages only.

Relevant legal cases

Two cases offer some guidance on establishing extensions of time when there are concurrent events *Balfour Beatty Building v Chestermount Properties (1993)* and *Henry Boot Construction v Malmaison Hotel (Manchester) (1999)*. In the latter case the judge said:

> It is agreed that if there are two concurrent causes of delay, one of which is a relevant event and the other is not, then the contractor is entitled to an extension of time for the period of delay caused by the relevant event, notwithstanding the concurrent effect of the other event. Thus to take a simple example, if no work is possible on site for a week, not only because of exceptionally inclement weather [a relevant event], but also because the contractor has a shortage of labour [not a relevant event], and if failure to work during that week is likely to delay the works beyond the completion date by one week, then if he considers it fair and reasonable to do so, the architect is required to grant an extension of time of one week.

The *Malmaison* case was considered and His Honour Judge Seymour QC gave further support to this approach in the case of *The Royal Brompton Hospital NHS Trust v Frederick Alexander Hammond (No 7) (2001)*. In a more recent case of *Motherwell Bridge Construction v Micafil Vakuumtecchnik (2002)* when considering concurrent events, His Honour Judge Toulim QC agreed that his approach should follow the Henry Boot judgement. He commented:

> Crucial questions are (a) is the delay on the critical path? And if so, (b) is it caused by Motherwell Bridge? If the answer to the first question is yes and the second is no, then I must assess how many additional working days should be included.

Judge Toulim departed slightly from the guidance in the Henry Boot case and went onto say:

> other delays caused by Motherwell Bridge (if proved) are not relevant, since the overall time allowed for under the contract may well include the need to carry out remedial works or other contingencies. These are not relevant events, since the court is concerned with considering extensions of time within which the contract must be completed.

Judge Toulim commented that the approach must always be tested against an overall requirement that the result accords with common-sense and fairness.

It is quite common for delays to occur where both the contractor and the employer, or engineer or architect acting on his behalf, are causing delays both of which are occurring at the same time. The courts in the USA have had occasion to deliberate on this question. A situation may arise where the contractor is in delay, let us say, due to essential materials not arriving on programme while at the same time they are waiting for details from the engineer necessary to fix the materials when they arrive on site.

A widely employed legal maxim, which would be applicable to these circumstances, is one which states that *a party to a contract is not entitled to benefit from its own errors*. This being the case the employer would be prevented from deducting liquidated damages and the contractor from claiming additional payment.

Roger Knowles offers the following advice on concurrent delays: 'Bearing in mind the various theories previously explained the best advice one can offer is to suggest that the contractor selects the theory which best suits his case and to argue it as forcefully as possible' (Knowles, 1992). This would seem sound advice on many of the issues concerning claims.

15.6   Proving the delay

There are basically four commonly used techniques in order to prove the entitlement to a delay (Lane, N., 2005, 2006).

1. *As planned versus as built*: This is the most simplistic technique, which involves comparing the planned sequence and timing of the project with the actual sequence and timing. It does not require a critical path programme or separate the events or make any allowance for the contractor's inefficiency.
2. *As planned impacted*: This technique takes the contractor's initial planned programme then adds the delays for which the employer is responsible. In theory the contractor should be entitled to an extension of time for their effect and will themselves be responsible for the difference between the impacted finish date and the actual finish date. This approach is highly theoretical and may bear no relationship to what the contractor did on site.
3. *Time-impact analysis*: This technique takes a snapshot looking at the effect of the delay on the planned programme at the time the event occurs. The planned programme obviously needs to take account of the progress at the time the delay occurs, with the effects of the events then plotted on an updated planned programme. The disadvantage of this approach is that the snapshot approach may not embrace significant factors occurring between the snapshots.
4. *As built but for analysis*: This approach involves identifying the *actual* sequence of the works. Events that are the employer's risk under the contract are identified and extracted from the as-built programme to show how long the work would have taken but for the events at the employer's risk.

The main problems with using any of these approaches are that there is no consensus on the most suitable approach. The Society of Construction attempted to introduce some conformity by recommending the use of approach 3, that is, time-impact analysis. In practice however, it seems that the experts cannot reach a consensus on which approach is the most appropriate (Lane, 2006).

A simple time-impact analysis will require an approach that takes into account the following:

1.   as planned network validated;
2.   known or notional employer delays added into network to model effect on programme;
3.   network time analysed to calculate revised completion period;
4.   amount by which revised completion period extends beyond due date is extension of time (EoT) entitlement.

This approach has the following advantages: relatively cheap and easy to prepare, and easy to agree between parties or between experts. However, it has the following disadvantages: theoretical, takes no account of actual methods and sequences of construction, nor of actual progress; unlikely to be accepted as proof by tribunal and does not assist with concurrency of delays.

By contrast, a sophisticated time-impact analysis will use the programme current at the time of delay. It will have the following advantages: takes into account contractor's progress up to the time of delay; takes into account the contractor's intended planning at the time of the delay; less theoretical assessment of EoT entitlement as at the time of the delay.

However the sophisticated approach will have the following disadvantages:

very difficult for tribunal to verify and hence trust the results. If the information being used is not correct then the results will prove nothing – 'garbage in – garbage out'.

(Marshall, 2005)

15.7   Disruption

The concept of delay, that is, lateness, is readily understood; disruption on the other hand is more complex. Disruption causes loss of production, disturbance and hindrance and could be one of three kinds:

1.   The work in question takes longer to complete, using the original resources;
2.   The work takes the same time because of increased resources;
3.   The contract takes the same time to complete, but certain resources are kept on site longer than originally necessary.

Thus it can be seen that the contractor may be entitled to additional costs for disruption even though he has completed the contract on time or within the extended period.

The SCL Protocol (2002) identifies that the most appropriate way to establish disruption is to apply a technique known as *the Measured Mile*. This compares the productivity achieved on an un-impacted part of the contract with that achieved on the impacted part.

15.8   Progress records

The contractor and architect/engineer commonly keep these records. If possible the records should be taken jointly or agreed/disagreed at the time of compilation. Well-maintained and accurate records form the backbone of most successful claims.

The following should be included as a realistic minimum:

•   A master programme based on a critical-path network, together with subsequent updates. It is important that programmes showing progress at a certain time should be saved rather than overwritten with the progress of the following period;

Table 15.1 Delay/disruption schedule.

Ref no.	Cause and effect of delay or disruption	Period of delay to sections or part of works	Period of delay to completion date	Contract clause relevant to delay	Contract clause relevant to loss/expense or contract clause breached	Date of delay notices and particulars	Date of loss and expense notice and particulars
1	Introduction of additional piling to S.E. corner of Block 'A' Letter 29.07.08 refers	3 days	3 days	2.29.1	4.24.1	Contractor's letters 29.07.08 03.09.08	Contractor's letters 29.07.08 03.09.08

- Records of progress achieved and labour and plant resources applied;
- Labour allocation sheets showing where the operative is working and when;
- Plant records showing when plant is working and when it is standing;
- Progress photographs/video records;
- Site diaries in standard format – a daily record of the job in progress;
- A drawing register kept up-to-date as new drawings are issued;
- Payroll records showing overtime worked and production records during these periods;
- Handwritten notes taken at meetings;
- Emails;
- Details of target-cost or bonus system operating;
- A weekly log of activities commenced, completed and problematic;
- Budgeted and actual costs and man-hours;
- Compilation of standard delay and disruption schedules (similar to *Scott Schedule* used in litigation/arbitration – see Table 15.1).

15.9 Claims presentation

The contractor should send to the architect/contract administrator/engineer as soon as possible their *notice of claim* which should

1. explain the circumstances giving rise to the claim;
2. explain why the contractor considers the employer to be liable;
3. state the clause(s) under which the claim is made.

The contractor should, as soon as possible, follow up this *notice of claim* with a detailed *submission of claim*, which should contain the following:

1. A statement of the contractor's reasons for believing that the employer is liable for extra cost with reference to the clauses under which the claim is made;
2. A statement of the event giving rise to the claim, including the circumstances that he could not reasonably have foreseen;

3.   Copies of all relevant documentation, such as:

    a.   contemporary records substantiating the additional costs as detailed;
    b.   details of original plans in relation to use of plant, mass-haul diagrams involved;
    c.   relevant extracts from the tender programme and make of major BofQ items;
    d.   information demonstrating the individual or cumulative effect of site instructions, variation orders and costs relating to the claim.

4.   A detailed calculation of entitlement claimed, with records and proofs.

A contractor's claim should be submitted in a similar format to that required for a *statement of case* in the courts. It should be self-explanatory, comprehensive and readily understood by someone not connected with the contract. It should contain the following sections:

1.   Title page
2.   Index
3.   Recitals of the contract particulars
4.   Relevant clauses and reasons for the claim
5.   Evaluation
6.   Appendices.

15.10   Quantifying the claim

The objective of all claims is to put the contractor back into the position he would have been in but for the delay; the original profit (or loss) should remain as included in the bid. It is therefore necessary to consider the *actual* additional costs incurred by the contractor at the time of the loss – provided of course that such costs have been reasonably incurred. It can be appreciated therefore that basing the evaluation on the contractor's tendered preliminaries is incorrect – even though this is the method sometimes used in practice for expediency.

The items described under the headings below are frequently encountered as *heads of claim*.

On site establishment costs

These are often called site overheads or simply *preliminaries* because the prices are found in the preliminary section of the BofQ. However all these costs should be ascertained from the contractor's cost records – these are the equivalent of damages at common law. It is also noted that the site establishment should be recorded when delay occurred and not at the end of the project when the resources will be running down.

Overheads are established based on contractor's contemporary records including the following:

• Supervisory and administrative staff;
• Site accommodation, including welfare and toilets;
• Construction equipment and tools for example tower cranes, scaffolding etc.;
• Site services, telephones, electricity.

Head office overheads

In principle, head office overheads are recoverable, however difficult to ascertain in practice. The contractor should make all reasonable efforts to demonstrate through records the head office overheads that it has failed to recover. If not feasible then the following formulae may be used with caution.

Hudson formula

The formula appears on p. 1076 of Hudson's *Building and Engineering Contracts*:

$$\frac{h}{100} \times \frac{c}{cp} \times pd$$

where h is the head office overheads and profit included in the *contract*, c is the contract sum, cp the contract period in weeks and pd the period of delay in weeks.

Emden formula

An alternative is produced in Emden's *Building Contracts and Practice*, vol. 2, p. 46

$$\frac{h}{100} \times \frac{c}{cp} \times pd$$

where h is the head office percentage arrived at by dividing the *total* overhead cost and profit of the contractor's organization as a whole by the *total* turnover, c the contract sum, cp the contract period in weeks and pd the delay period in weeks.

Eichleay formula

This formula is best known and most widely used in the US. The formula computes the daily amount of overhead that the contractor would have charged to the contract had there been no delay. The formula is developed in three stages:

Stage 1

$$\frac{\text{Contract billings}}{\text{total billings for actual contract period}} \times \begin{array}{l}\text{total H.O. overhead incurred} \\ \text{during contract period}\end{array} = \begin{array}{l}\text{overhead} \\ \text{allocatable} \\ \text{to the contract}\end{array}$$

Stage 2

$$\frac{\text{Allocatable overhead}}{\text{actual days of contract performance}} = \text{Head office overhead allocatable to contract per day}$$

Stage 3

$$\begin{array}{l}\text{Overhead allocatable} \\ \text{to contract per day}\end{array} \times \begin{array}{l}\text{number of days of} \\ \text{compensable delay}\end{array} = \begin{array}{l}\text{unabsorbed} \\ \text{overhead}\end{array}$$

In the cases of *Alfred McAlpine Homes North v Property and Land Contractor (1995)* and *Amec Building Limited v Cadmus Investments (1996)* the court simply calculated the contractor's average weekly costs (by reference to the company's accounts) multiplied by the number of weeks of delay and then allocated to the particular contract by means of a pro-rata calculation based upon the value of the work carried out on the site during the overrun period and the value of all works being carried out by the contractor during the overrun period.

In his decision of *JR Finnegan Ltd v Sheffield City Council (1988)* Judge William Stabb said:

It is generally accepted that, on principle, a contractor who is delayed in completing due to the default of his employer, may properly have a claim for head office off-site overheads during the period of delay on the basis that the work-force, but for the delay, might have had the opportunity of being employed on another contract which would have had the effect of funding the overheads during the overrun period.

Chappell (1998) identifies that there is a use for formulae in appropriate situations, usually as a last resort, where it is clear there has been a loss, but where there is a complete lack of proper evidence. However, their uncritical use without regard to available facts and without supporting evidence is not recommended.

Interest and financing charges

In order for reimbursement to be made under this heading loss must be actually suffered. Further to the cases of *F.G. Minter Ltd v Welsh Health Technical Services Organisation (1980)* and *Rees & Kirby Ltd v Swansea City Council (1985)* it is now evident that contractors can recover as *cost* or *direct loss and expense* either

- the interest payable on the capital borrowed, or
- the interest on capital that would otherwise have been invested.

This approach was confirmed as valid in the case of *Amec Process & Energy v Stork Engineers & Contractors (2002)*. Amec put forward a claim for interest on alternative footings. First, compound interest was claimed on the basis that the contract terms allowed for the reimbursement of financing charges. Second, interest was claimed as damages for breach of contract. Finally, as a fall-back position, Amec claimed statutory simple interest pursuant to the court's discretionary powers to award interest on judgement sums. Amec had financed the work from its own resources and from the use of inter-company loans and financing facilities provided by its parent company.

Judge Thornton concluded that, having regard to the valuation provisions of the contract, nothing could be fairer or more reasonable than that the extra cost to Amec of funding the additional work should be recoverable from Stork. Following *Rees v Kirby*, Amec was also entitled to the calculation of interest on a compounded basis to reflect the manner in which its works were financed.

It followed that the claim for financing could also be made as one for damages for breach of contract pursuant to the second limb of the *Hadley v Baxendale (1854)* case, being a foreseeable loss in specific knowledge of the parties at the time of making the contract.

The appropriate rate of interest is normally based on the actual rate paid by the contractor provided it is not excessive. Financing charges should be calculated using the same rates and methods as the contractor's bank; for example, compounding interest at regular intervals.

Increased costs

If the contract is fixed price then the additional cost of carrying out the work later than anticipated due to delay and disruption is generally recoverable. Normally reimbursement would be on the basis of a known formula, for example, NEDO or *Baxter*.

Profit

Loss of profit that the contractor would have earned but for the delay and disruption is an allowable head of claim following the rule established in the *Hadley v Baxendale (1854)* case. However in order to succeed in such a claim the contractor must be able to prove that he has been prevented from earning profit elsewhere.

Loss of productivity/winter working

Inefficient use of labour and plant is an acceptable head of claim. It can be established by comparing the production rates during the disrupted period with those rates achieved prior to the disruption.

Costs of claim preparation

Again the contractor must be able to prove that they have incurred additional costs. Leading commentators seem to concur that the contractor's additional cost in preparing the claim and/or the cost of outside consultants is recoverable provided that the item has not been claimed elsewhere, for example, site overheads.

In the case of *Richards and Wallington (Plant Hire) Limited v Devon County Council (1984)*, the costs of the claimant's staff who were not acting as experts was disallowed as a recoverable cost. However this approach was overturned in the recent case of *Amec Process & Energy v Stork Engineers & Constructors (2002)*. Amec had engaged its own personnel in collating, analysing and presenting the primary and supporting evidence to be used by its expert witness. His Honour Judge Thornton QC noted that the Civil Procedures Rules defined recoverable costs as including *fees, charges, disbursements, expenses and remuneration*. Judge Thornton was satisfied that the time charges incurred by Amec in employing its own personnel fell within each of these categories of cost.

Global claims

The SCL Protocol (2002) identifies that the practice of contractors making composite or global claims without substantiating cause and effect is discouraged and rarely accepted by the courts.

In general, it is necessary for the contractor to establish each and every head of claim, by means of supporting documentation and other evidence. The global approach was recognized in *J. Crosby & Sons Ltd v Portland UDC (1967)* which was decided under the ICE *Conditions of Contract* 4th edition. However this approach should be the exception not the rule, only applicable where numerous/complex/interrelated issues. Doubt was cast on the global approach following the Privy Council's decision in the Hong Kong case of *Wharf Properties Ltd v Eric Cumine Associates (1991)* where the client's action against their architect for negligent design and contract administration were struck out as incomplete and therefore disclosing no reasonable course of action.

Following the case of *How Engineering Services Ltd v Lindner Ceiling Partitions plc (1995)*, Chappell (1998) considered that the courts have clearly set out what is required of a contractor when making a claim.

- The claimant must set out an intelligible claim, which must identify the loss, why it has occurred, and why the other party has an enforceable obligation recognized at law to compensate for the loss.
- The claim should tie the breaches relied on to the terms of the contract and identify the relevant contract terms.

- Explanatory cause and effect should be linked.
- There is no requirement that the total amount for the loss must be broken down so that the sum claimed for each specific breach can be identified. But an all-or-nothing claim will fail in its entirety if a few causative events are not established.
- Therefore a global claim must identify two matters:

  a.   The means by which the loss is to be calculated if some of the causative events alleged have been eliminated. In other words, what formula or device is put forward to enable an appropriate scaling down of the claims to be made?
  b.   The means of scaling down the claim to take account of other irrevocable factors such as defects, inefficiencies or events at the contractor's risk.

The case of *John Doyle Construction v Laing Management (Scotland) Limited (2004)* provided an important reassessment of global claims. Stephen Furst QC (2005) Joint Editor of *Keating on Building Contracts* considers that this case has resulted in three main changes in the emphasis of the law.

- Whereas previously it was understood that *any* cause of loss shown not to be the responsibility of the defendant would be fatal to the global claim, it now appears that this only applies if the cause of loss is *significant or dominant*.
- The court seemed comfortable with the idea of apportionment of loss by the tribunal between causes for which the employer is not liable, even if this may be a rough and ready process.
- The issue of whether causation can be proved should normally wait until the trial when all the evidence is in and so, presumably, would not be decided at the interlocutory stage on an application to strike out.

15.11   Conclusion

Claims submissions are inevitable on construction projects. Delays will be caused to the project which are outside the control of the contractor. These may entitle the contractor to additional costs as well as EoT. This chapter has identified some of the key issues and demonstrated the legal and administrative complexity of the subject. The parties will need to be skilled negotiators in order to avoid a lengthy arbitration or a court case.

From the employer's viewpoint, claims settled early are usually settled cheaply, for contractors will seldom be able to anticipate the full impact of delay and disruption until receipt of sub-contractors' and suppliers' final invoices.

15.12   Some legal cases

Below listed are some of the more important cases involving construction claims:

Interest and financing charges

Ogilvie Builders Ltd v Glasgow City District Council (1994) [Court of Session] 68 BLR 122, (1994) CILL 930, 41 Con LR 1
Blaenau Gwent Borough Council v Lock (Contractors Equipment) Ltd (1994) [OR] 71 BLR 94 (1993) CILL 904, 37 Con LR 121
Kingston-upon-Thames (Royal Borough) v AMEC Civil Engineering Ltd (1993) 35 Con LR 39

Secretary of State for Transport v Birse-Farr Joint Venture (1993) [QBD] 62 BLR 36, (1993) CILL 903, (1993) 35 Con LR 8, (1993) 9 Const LJ 213

Costain Building & Civil Engineering Ltd v Scottish Rugby Union plc (1994) [Court of Session, Inner House] 69 BLR 80, 43 Con LR 16

Farrans (Construction) Ltd v Dunfermline District Council (1988) [Court of Session – Inner House] 4 Con LJ 314

Holbeach Plant Hire Ltd and Another v Anglian Water Authority (1988) [OR] (1988) CILL 448, 14 Con LR 101

Royal Borough of Kingston upon Thames v AMEC Civil Engineering Ltd (1993) [OR]

Morgan Grenfell (Local Authority Finance) Ltd v Seven Seas Dredging Ltd (No 2) (1990) [OR] 51 BLR 85, (1990) CILL 618, 21 Con LR 122, (1991) 7 Const LJ 110

Amec Process & Energy v Stork Engineers & Contractors BV (No 4) (2002) [QBD (TCC)] CILL 1883

Extensions of time and liquidate and ascertained damages

Balfour Beatty Building Ltd v Chestermount Properties Ltd (1993) [Commercial Court] 62 BLR 1, 32 Con LR 39, (1993) 9 Const LJ 117

Fairweather (H) and Co Ltd v London Borough of Wandsworth (1987) [OR] 39 BLR 106

John Barker v Portman Hotel (1996) 83 BLR 31

Peak Construction (Liverpool) Ltd v McKinney Foundations Ltd (1970) [CA] 1 BLR 11, (1970) 69 LGR

Temloc Ltd v Errill Properties Ltd (1987) [CA] 39 BLR 30, (1986/87) CILL 376, 12 Con LR 109, (1988) 4 Const LJ 63

Anson Contracting Limited v Alfred McAlpine Construction Isle of Man (1999) TCC

Balfour Beatty Construction Ltd v The London Borough of Lambeth (2002) TCC

The Royal Brampton Hospital NHS v Watkins Gray International (UK) (2000) TCC

Concurrent events

Motherwell Bridge Construction Limited v Micafil Vakuumtecchnik (2002) TCC 81 Con LR 44

H Fairweather & Co v London Borough of Wandsworth (1987) 39 BLR 106

Balfour Beatty Building v Chestermount Properties (1993) 62 BLR 1

Henry Boot Construction v Malmaison Hotel (Manchester) (1999) 70 Con LR 32

The Royal Brompton Hospital NHS Trust v Frederick Alexander Hammond (No 7) (2001) 76 Con LR 148

Motherwell Bridge Construction v Micafil Vakuumtecchnik (2002) TCC 81 Con LR 44

Acceleration

Glenlion Construction Ltd v The Guiness Trust (1987) [OR] 39 BLR 89, (1986/7) CILL 360, (1998) 4 Const LJ 39

John Barker Construction Ltd v London Portman Hotel Ltd (1996) 83 BLR 35

Anson Contracting Limited v Alfred McAlpine Construction Isle of Man Limited (1999)

Amec & Alfred McAlpine (Joint Venture) v Cheshire County Council (1999) BLR 303

Global claims

Crosby (J) and Sons v Portland Urban District Council (1967) [OR] 5 BLR 121

Wharf Properties Ltd v Eric Cumine Associates (No 2) (1991) [PC] 52 BLR 1, (1991) CILL 661, 29 Con LR 113, (1991) 7 Const LJ 251

British Airways Pension Trustees Ltd v Sir Robert McAlpine & Sons Ltd and Others (1994) [CA]
Mid-Glamorgan County Council v J Devonald Williams and Partner (1991) [OR] (1992) CILL 722, 29 Con LR 129, (1991) 8 Const LJ 61
John Doyle Construction Ltd v Laing Management (Scotland) Ltd (2002)
John Doyle Construction Limited v Laing Management (Scotland) Limited, Inner House Court of Session, 11 June 2004

Engineer's duty to contractor/tenderer and when preparing bills

Christiani & Neilsen Ltd v Birmingham City Council (1994) [OR] [1995] CILL 1014
Pacific Associates Incorporated and RB Construction Ltd v Baxter and Others (1988) [CA]
Blackpool and Fylde Aero Club Ltd v Blackpool Borough Council [1990] 1 WLR 1195, [1990] 3 All ER 25, 88 LGR 865

Unforeseeable conditions under clause 12 of the ICE Conditions/quality of Materials

Humber Oils Terminal Trustees Ltd v Hersent Offshore Ltd (1981) 20 BLR 22
Humber Oils Terminal Trustee v Harbour & General Works (1991) [CA] 59 BLR 1
Rotherham Metropolitan Borough Council v Frank Haslam Milan and Company Ltd and MJ Gleeson (Northern) Ltd (1996) [CA] 78 BLR 1, (1996) 12 Const LJ 333
Young & Marten Ltd v McManus Childs Ltd (1969) [HL] 2 All ER 1169, [1968] 3 WLR 630
Christiani Neilsen v Bachy Ltd [ORB]; 16 June 1995

15.13   Questions

1.   A building contractor intends to submit to the client a claim for loss and expense because of late delivery of drawings (JCT 05 with *Quantities* applies). Outline the factors you would need to consider in compiling such a claim and discuss what information you would expect to provide in order to support your claim.

2.   A contractor (C) is awarded a £10 million contract to construct a 25 km pipeline over hilly country. The commencement date is anticipated to be the first week in May 2007 with a time for completion of 18 months.

   The Contractor submitted his clause 14 programme, which was approved by the engineer, showing a close down during January, February and March because of anticipated bad weather.

   The Contractor was delayed by:

   a.   failure by the engineer to provide the drawings necessary for the contractor to construct the work, on the part of the site most inaccessible to plant, in time to meet the programme (eight-week delay – June/July 2008);

   b.   failure by a nominated supplier (S) to deliver the pipes required in accordance with the programme agreed between C, S and the engineer (four-weeks delay – April/May 2008) the ground conditions proving more difficult than expected (eight-week total September/October 2007);

   c.   it is proving harder than expected to engage suitably skilled operatives to undertake the work (four-week delay throughout project);

   d.   adverse weather in December 2007 (from 15 to 31);

   e.   restricted access to the site due to an outbreak of Foot and Mouth in the local cattle (12-week delay – April to June 2008);

f.   breakdown of pipe-laying equipment (two weeks in October 2007);

g.   national strike by ready-mix concrete suppliers (one week in November 2007).

You have been appointed by the contractor to assist them with their claim – they have claimed 41-weeks EoT all with associated costs. Write to advise the contractor how they should proceed to obtain additional payment from the employer for their additional costs in each case, and indicate the likelihood of success; clearly identify the relevant clauses in the ICE *Conditions of Contract*.

You can base your solution on either the ICE *Conditions of Contract*, 7th edition or the FIDIC 1999 form.

3.   Prepare a brief written summary (no more than 250 words) of TWO cases involving construction claims clearly identifying the issues involved. Make an oral presentation to your peers on your findings.

Bibliography

Burr, A. and Palles-Clark, R. (2005) 'The consideration of critical path analysis in English Law', *Construction Law Journal*, vol. 2, no. 3, pp. 222–241

Chappell, D. (1998) *Powell-Smith and Sims' Building Contract Claims*, 3rd edition, Blackwell Science

Furst, S. (2005) 'Global Claims in 2005 (UK)', lecture given on Tuesday 21 June 2005 www.rics.org/NR/rdonlyres/B562D6B4-6600-4565-885E-E563465110D5/0/GlobalClaimsin2005.pdf (accessed 16 April 2007)

Furst, S. and Ramsey, V. (2001) *Keating on Building Contracts*, 7th edition, Sweet & Maxwell

Knowles, R. (1992) *Claims – Their Mysteries Unravelled: An Introduction to Claimsmanship for Contractors and Subcontractors*, JR Knowles

Knowles, R. (2001) 'Hudson's formula revisited – a recent case', *Civil Engineering Surveyor*, June, pp. 14–15

Lane, N. (2005) 'Listen, this is important', *Building* magazine, 2 December

Lane, N. (2006) 'How to be good', *Building* magazine, 17 February

Linnett, C. and Lowsley, S. (2006) 'A simple approach', *The Journal RICS Construction*, November/December, RICS

Marshall, J. (2005) 'Delay Analysis', lecture to MSc Construction Law/MSc Construction Project Management students, University of Wolverhampton

Pickavance, K. (2006) 'A case for the defence', *The Journal RICS Construction*, September

Thomas, R. (1993) *Construction Contract Claims*, Macmillan

Trickey, G. and Hackett, M. (2001) *The Presentation and Settlement of Contractors Claims*, EF&N Spon

Useful websites

www.jrknowles.comwww.atkinson-law.com

www.brewerconsulting.co.uk

www.scl.org.uk *The Society of Construction Law Delay & Disruption Protocol* (2002)

The Protocol is not intended to be a contract document, but it does provide recommendations and guidance to those involved with drafting contracts. It is intended to act as an aid in the interpretation of delay and disruption disputes on standard form build and civil engineering contracts.

Part VI
Contracts and case study

16 The NEC *Engineering and Construction Contract*

16.1 Introduction

Currently used on over 6,000 contracts world-wide from small projects to large internationally known projects such as the Channel Tunnel Rail Link, the NEC *Engineering and Construction Contract* (NEC ECC) has established itself as the number one form of contract that helps avoid disputes, delays and ultimately extra costs.

Widely adopted in the civil engineering sector, the NEC is now making major inroads into the building sector. Among the many notable users is the Highways Agency, The Environment Agency, BAA, Sainsbury's, London Underground, NHS Estates – through its £2 billion ProCure21 framework programme for hospital buildings, BT and the Welsh Assembly. The majority of UK local authorities use the NEC for highway and drainage projects and increasingly for schools, offices and social housing schemes. In 2006 it was announced that the UK Olympic Delivery Authority (ODA) had chosen the NEC3 suite of contracts to procure all fixed assets and infrastructure for the London 2012 Olympic Games.

The NEC ECC is radically different from other standard forms of contract currently in use. Introduced by the Institution of Civil Engineers (ICE) in 1993, the form has great flexibility and can be used on all types of construction and engineering works. It aims to be easier to understand and use than conventional contract forms and stimulate effective project management by encouraging a *foresighted co-operative approach*. The philosophy and objective of the NEC is to create an open, cooperative, no-blame, non-adversarial team approach to managing contracts.

The launch of the NEC 3rd edition in July 2005 coincided with a unique endorsement from the UK Office of Government Commerce (OGC). The endorsement reads:

> OGC advises public sector procurers that the form of contract used has to be selected according to the objectives of the project, aiming to satisfy the Achieving Excellence in Construction (AEC) principles. This edition of the NEC (NEC3) complies fully with the AEC principles. OGC recommends the use of NEC3 by public sector construction procurers on their construction projects.

16.2 The NEC family of contracts

The NEC3 comprises a family of standard integrated contracts incorporating the following standard forms:

- The NEC3 *Engineering and Construction Contract* (ECC) – between the employer and contractor for construction and engineering works;

Panel 16.1 Key differences between NEC2 and NEC3

The 3rd edition of the NEC ECC was published as part of the revised extended suite of contracts on 14 July 2005.

The key differences between NEC 2nd edition and NEC3 are:

- Choice of two adjudication clauses, including a Construction Act-compliant procedure;
- Optional key performance indicators for bi-party contracts;
- Optional limitation of liability clauses;
- Introduction of risk-reduction meetings and risk register;
- Introduction of key dates;
- New prevention clause with associated compensation event and termination provisions;
- Compensation event quotations accepted if project manager does not reply;
- Swift compensation event assessment for options A and B;
- Options C, D and E monthly assessment includes forecast to next assessment;
- Simplified schedule of cost components;
- Two fee percentages introduced.

Source: NEC Users' Group Newsletters – No. 33, June 2005, and for an in-depth analysis of see Robert Gerard's review in No. 34 dated September 2005.

- The NEC3 *Engineering and Construction Short Contract* (ECSC) – between the employer and contractor, for low-risk, straightforward work;
- The NEC3 *Term Services Contract* (TSC) – for engaging suppliers of services for a period of time;
- The NEC3 *Framework Contract* (FC) – for engaging suppliers to provide services operating under a framework;
- The NEC3 *Engineering and Construction Subcontract* (ECS) – between a contractor and subcontractor, back to back with ECC;
- The NEC3 *Engineering and Construction Short Subcontract* (ECSS) – between a contractor and subcontractor, under either ECC or ECSC;
- The NEC3 *Professional Services Contract* (PSC) – for engaging a supplier of professional services;
- The NEC3 *Adjudicator's Contract* (AC) – for engaging an adjudicator to settle disputes between parties under an NEC contract.

The contracts are supported by guidance notes, flow charts and the advisory NEC3 *Procurement and Contract Strategies*.

The chart in Fig 16.1 demonstrates how a design and build (bi-party partnering) contract is administered using the NEC family of contracts.

16.3 Objectives of the NEC

The NEC is a simple and flexible form of contract designed around a concept of common purpose. The primary objective is to shift the emphasis of control from procedures for the calculation of extra payment to the contractor, if things go wrong, to arranging matters so that things are less likely of going wrong.

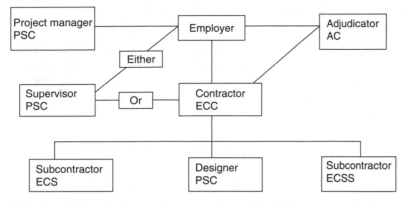

16.1   The NEC family of contracts under design and build (bi-party partnering).

Clarity

The form is written in plain English and long sentences have been avoided where possible. There is a simple structure and clause numbering. A flowchart has been developed as a check on drafting and as an aid for users.

The actions by parties are defined precisely in the contract and the time periods for all-important actions are deliberately set tight to motivate timely responses. Subjective phrases such as *fair* and *reasonable* have been avoided. Moreover, there is a single procedure for assessing all compensation events, which includes an assessment of both cost, and time for all events; it is not retrospective and once carried out is fixed.

Flexibility

The NEC is intended to be used on all types of engineering, construction and mechanical and electrical work. It allows the degree of contractor design responsibility to be varied from 100–0% – the extent is defined in the Works Information. The contract is also adaptable for use on international contracts.

The NEC provides a choice of tender and procurement arrangements to suit most project circumstances. This allows the allocation of risks to suit the particular contract through the use of priced contracts, target contracts, cost-reimbursable contracts and management contracts.

The NEC provides nine core clauses, which are common to all options with six main payment options, and secondary options allowing the user to fine-tune the risk allocation.

Stimulus to good project management

The NEC is founded on three key principles:

Principle 1 – foresighted, cooperative management shrinks risks and mitigates problems

The NEC3 achieves this by including: an *early warning* procedure for identifying future problems and minimizing their impact; a regularly updated and agreed programme with method statements and resources showing timing and sequencing of employer and contractor actions; assessment of time and cost as the contract progresses ideally before the work is executed and stated maximum timescales for the actions of parties.

*Principle 2 – both parties are motivated to work together if it is in their
professional and commercial interests to do so*

The NEC3 achieves this by including a clear statement of events for which the employer is liable
(compensation events) and a structured method of calculating changes in contractor's costs.

Furthermore there are sanctions on the contractor to early warn; submit a first programme
containing the information required; maintain an up-to-date Accepted Programme and to
provide realistic and timely quotations.

*Principle 3 – clear division of function and responsibility helps accountability
and motivates people to play their part*

The NEC3 achieves this by clearly identifying the roles of the key players. The traditional role of
the engineer or architect has been split into four roles.

Role 1 – the designer (not mentioned in the contract): The designer's role is to develop the
design to meet the employer's objectives to the point where tenders for construction are to be
invited. If a *design and construct* contract is envisaged, the employer's designer's role is restricted
largely to providing a *performance specification* together with standards for design and materials,
which they may wish to specify for inclusion in the works information.

Role 2 – the project manager: The project manager is responsible for time and cost
management on the employer's behalf together with management of the designers' activities.
The project manager is appointed by the employer, either from their own staff or from outside
and is the spokesperson for the employer.

Role 3 – the supervisor: The supervisor is responsible for ensuring that the quality of
construction meets that in the works information/specification. The role of the supervisor is similar
to that of a resident engineer or architect, who may be assisted by an inspector or clerk of works.

Role 4 – the adjudicator: The adjudicator is an independent third party brought in to rapidly
resolve any disputes. The adjudicator becomes involved only when a dispute is referred to them.
If either party does not accept their decision, they may proceed to the tribunal (either arbitration or the courts). Under the adjudicator's contract, payment of the adjudicator's fee is shared
equally by the parties.

16.4   Design principles

In order to create a contract under the NEC3 the employer is required to specify and include the
following:

- one of the six main payment options;
- the nine core clauses;
- one of the dispute resolution options;
- which of the 17 secondary options apply – if any;
- any additional clauses under secondary option Z.

Main payment options

The main options are the following contract types, one of which must be chosen:

Option A – priced contract with activity schedule: this is a lump-sum contract based upon a
priced-activity schedule submitted at tender stage. This option is appropriate for use when the

employer designs the work in the form of drawings and specification or where a performance specification is provided by the employer.

The tenderer decides how to break up the work into activities, enters them on the schedule and prices each one as a lump-sum item. If the employer wants to specify particular activities, which the contractor is to identify in the activity schedule, they should state their requirements in the instructions to tenderers.

An activity is essentially one bar on a bar chart, which the contractor has priced. Activities can include fixed overheads, time-related charges, temporary works, materials on site as well as the permanent works. The activity schedule is extracted from the programme, but is a separate document.

Option B – priced contract with bill of quantities (BofQ): This is a lump-sum contract based upon a priced BofQ submitted at tender stage. This option is used where the project is well defined at tender and designed by the employer's design consultants but some changes are anticipated.

Option C – target contract with activity schedule: This is a cost-reimbursable contract, which is awarded to the contractor on the basis of a priced-activity schedule, prepared by the contractor and submitted with their tender. Payment for carrying out the works is on a cost-reimbursable basis with an additional amount paid if the final cost is less than an agreed target cost, which is adjusted for compensation events.

Option D – target contract with BofQ: This is a cost-reimbursable contract, which is awarded to the contractor on the basis of a priced BofQ, normally prepared by the employer, submitted at tender stage. Payment for carrying out the works is on a cost-reimbursable basis with an additional amount paid if the final cost is less than the agreed target cost, which is adjusted for compensation events and remeasurement.

Options C and D are used where the work is not fully defined or where the anticipated risks are greater or where the employer sees a significant benefit in encouraging collaboration through the target mechanism.

Option E – cost-reimbursable contract: This is a cost-reimbursable contract, which is awarded to the contractor on the basis of a fee percentage submitted at tender stage. This approach is suitable for refurbishment work, for enabling work or for emergency repair work.

Option F – management contract: This is a cost-reimbursable contract, which is awarded to the contractor on the basis of a fee percentage submitted at tender stage.

All the main options can be used with the boundary between design by the employer and design by the contractor set to suit the chosen strategy. If the works information set down by the employer is only a performance specification, most of the design will be done by the contractor (effectively a *design and construct contract*). If the works information includes detailed drawings and specifications, little design remains for the contractor to complete.

The ECC *Guidance Notes* recommend that the following factors be taken into account when selecting the contract strategy:

- Who has the necessary design expertise?
- Whether there is particular pressure to complete quickly?
- How important is performance of the complete works?
- Whether certainty of final cost is more important than lowest final cost?
- Where can a risk be best managed?
- What total risk is tolerable for contractors?

- How important is cross-contract coordination to achievement of project objectives?
- Whether the employer has good reasons for himself selecting specialist contractors or suppliers for parts of the work?

The result of these considerations should be a statement of the chosen contract strategy comprising the following:

- A schedule of the parts of the project which will be let as separate contracts;
- For each contract there be a statement of the stages of work which includes management, design, manufacture, erection, construction, installation, testing and commissioning as appropriate;
- A statement of the ECC main option which will be used for the contract.

16.5   Core clauses

There are nine core clauses, all of which are common to all NEC ECC contract types:

1. General (sub-clauses 10–9)
2. The *Contractor's* main responsibilities (sub-clauses 20–27)
3. Time (sub-clauses 30–36)
4. Testing and defects (sub-clauses 40–45)
5. Payment (sub-clauses 50–52)
6. Compensation events (sub-clauses 60–65)
7. Title (sub-clauses 70–73)
8. Risks and Insurance (sub-clauses 80–87)
9. Termination (sub-clauses 90–93)

1. *General:* This clause defines the duties and obligations of the parties, procedures that are to be followed, communication methods, interpretation of the contract and early warnings that must be given. Clause 10 notes the requirements for the parties to act in a spirit of mutual trust and cooperation. Thus the NEC3 contract provides an ideal basis for a partnering arrangement.
    The risk register [clause 11.2(14)]: This is a new addition to NEC3. Initially it will contain those risks identified by the employer and the contractor in parts one and two respectively of the contract data. Risks are then added to the risk register as part of the early warning process described in clause 16, or removed because of actions taken by the parties to avoid them or because they did not happen.
    Early warning [clause 16]: The purpose of early warning is to make it compulsory for the project manager or the contractor to call an early warning meeting as soon as possible to discuss anything which may affect the cost, timing of completion and quality of the works. The sanction for failure by the contractor to give early warning is to reduce the payment due to them for a related compensation event [clause 63.5].
2. *Contractor's main responsibilities:* The contractor's main responsibilities are set out in this clause, including that of design. The contractor's basic obligation is to *provide the works* which is defined as including supplying all the necessary resources to achieve the end result including designing, fabricating, delivering to site, erecting, constructing, installing, testing and making good defects. The works information, defined in clause 11.2(19) provided by the employer, should state everything which is intended concerning the work, including design work, which should be done by the contractor.
    Under clause 21.5 the contractor's liability for his design can be limited to an amount stated in the contract data. This liability can also be limited to reasonable skill and care by inclusion of option X15.

3. *Time:* All matters concerned with time are dealt with by this clause, such as start and completion dates, programming, possession of the site and taking over of the works. The period of time within which the contractor is required to provide the works is not stated. Instead, the starting date and the completion date are given in the contract data. The programme is an important document for administering the contract as it enables the project manager and the contractor to monitor progress and to assess the time effects of compensation events including changes to the completion date.

    Clause 31.2 lists the information, which the contractor is required to show on each programme submitted for acceptance. This list of requirements is far more extensive than normally required under a traditional contract and includes such issues as float and time-risk allowances.

    Method statements for the contractor's operations are also required to be submitted, detailing construction methods as well as details of the resources, including equipment that they intend to use. Furthermore, any contractor's quotation submitted in relation to a compensation event which includes a revised programme must include any revised methods of construction and resources.

4. *Testing and defects:* This clause deals with searching and notifying defects, tests, correcting defects and uncorrected defects.

5. *Payment:* This clause deals with payment for work done, payment certificates and payment of actual cost. Payment mechanisms for the six main options (see Table 16.1) are based on the use of the three key terms: *The Prices* and *The Price for Work Done to Date* (PWDD) and the *Defined Cost*. The content of PWDD varies according to which main option is used.

    Clause 50.3 is designed to provide a powerful motivation on the contractor to submit a programme which contains the information required by the contract. If the programme is not submitted by the due date then one quarter of the price for work done to date should be retained.

6. *Compensation events:* This clause deals with compensation events, their identification, assessment and acceptance of quotations from the contractor.

    Compensation events are events which do not arise from the contractor's fault and may entitle the contractor to additional payment and also to additional time.

    Compensation events are listed in the core clauses, the options and the contract data. The main list is in the core clause 60.1 which includes compensation events (1) to (19). Events applicable to main options B and D are stated in clauses 60.4, 60.5 and 60.6. Other compensation events are stated in secondary option clauses X2.1, X14.2, X15.2 and Y (UK) 2.4.

Table 16.1 Payment mechanism for six main options.

Option	Prices	Price for the work done to date (PWDD)
A	Activity schedule prices for activities 11.2 (30)	Total of the prices for completed activities 11.2 (27)
B	BQ rates and lump sums 11.2 (31)	Quantities of completed work at BQ rates and proportions of lump sums 11.2 (28)
C	Activity schedule prices for activities 11.2(30)	Defined cost forecast to be paid before next assessment + Fee 11.2(29)
D	BQ rates and lump sums 11.2 (31)	As Option C
E	Defined Cost + Fee 11.2 (29)	As Option C
F	Defined Cost + Fee 11.2 (29)	As Option C

Source: NEC3 Guidance Notes 2005.

The ECC clause 60.1(12) on 'physical conditions' includes a procedure that is not radically different from the one found in traditional ICE *Conditions of Contract*.

By contrast, ECC clause 60.1(13) on 'adverse weather' introduces a completely innovative approach with compensation only being allowable if the weather conditions are more adverse than the weather conditions at the site location in the past ten years as proved by data produced by the UK Metrological Office.

Clause 61.6 considers the case where the nature of the compensation event may be such that it is impossible to prepare a sufficiently accurate quotation. One example of this is where unexpected physical conditions are encountered (compensation event 12) but their extent is unknown. In these cases, quotations are submitted on the basis of assumptions stated by the project manager in his instruction to the contractor. If the assumptions later proved to be wrong, the project manager's notification of their correction is a separate compensation event [clause 60.1(17)].

Under clause 63.1 assessment of compensation events as they affect prices is based on their effect on defined cost plus the fee. For options C, D and E, pricing of the various components of defined cost is normally based on the schedule of cost components (SCC) with the associated percentages tendered in the contract data part two, although the shorter SCC may be used in some circumstances (see clause 63.15). The shorter SCC is used with options A and B. The fee is calculated in accordance with clause 11.2(8). This approach is different from most standard forms where variations are valued using the rates and prices in the contract as a basis.

7. *Title:* The employer's title to equipment, plant and materials both off site and brought into the working area are dealt with by this clause.
8. *Risks and insurance:* This clause allocates the various defined risks between the parties and the insurance policies and cover they are required to provide.
9. *Termination:* This clause deals with the procedures if a dispute occurs, including adjudication, arbitration and termination.

Dispute resolution option

There are two options for dispute resolution:

Option W1 (used unless the Housing Grants, Construction and Regeneration Act 1996 applies) is the standard ECC Option and should be used for all contracts except those in the UK which are regulated by the HGC&R Act 1996.
Option W2 (used when the Housing Grants, Construction and Regeneration Act 1996 applies) should only be used in the UK where there is a 'construction contract' within the definitions in Sections 104 to 106 of the HGC&R Act 1996.

16.6   Secondary options

Note that all the secondary options are *optional* allowing the employer to build up the contract to suit their specific requirements:

Option X1: price adjustment for inflation – calculated based on a formula method enabling the employer to carry most of the risk of inflation;
Option X2: changes in the law – employer carries risk of changes in law after the contract date;
Option X3: multiple currencies – used in priced contracts only;
Option X4: parent company guarantee – required from the contractor often as an alternative to the performance bond;

Option X5: sectional completion – sections should be identified in contract data, part 1 prepared by the employer;

Option X6: bonus for early completion – additional motivation for contractor if early completion would benefit the employer;

Option X7: delay damages – these are the equivalent of liquidated damages which should be based on a genuine pre-estimate of the employer's likely losses;

Option X12: partnering – enables multiparty partnering agreement to be implemented;

Option X13: performance bond – form and amount should be stated in the contract data, part 1;

Option X14: advanced payment to the contractor – relevant if contractor will incur substantial 'up-front' costs for example in pre-ordering specialist materials plant or equipment; the amount must be stated in contract data, part 1;

Option X15: limitation of the contractor's liability for his design to reasonable skill and care – without this clause the standard liability for contractor's design is generally 'fitness for purpose'; this option reduces the liability to the standard of a consultant designing for the employer;

Option X16: retention – enables employer to retain a proportion of the price for works done to date until the contractor has completed the works. Then following a partial release the balance is released upon the issue of the defects certificate;

Option X17: low-performance damages – where performance in use fails to reach the specified level, the employer can take an action against the contractor to recover damages or as an alternative can recover liquidated damages under *Option X17*;

Option X18: limitation of liability – places limits on various liabilities of the contractor;

Option X20: key performance indicators – this option enables the contractor's performance to be monitored, this could be linked to the payment of bonuses if completed early or in case of exceptional accident record;

Option Y (UK) 2: The Housing Grants, Construction and Regeneration Act (1996);

Option Y (UK) 3: The Contracts (Rights of Third Parties) Act 1999;

Option Z: additional conditions of contract – this option enables further conditions to be added to the contract depending on the specific circumstances.

16.7 ECC tender documents

The documents to be issued with the invitations to tender include the following:

- Instructions to tenderers (including any instructions for preparing activity schedules under options A and C);
- A form of tender;
- Contract data, part 1;
- Contract data, part 2 (pro-forma for completion by tenderers);
- BofQ (options B and D);
- Works information;
- Site information;
- A pre-tender health and safety plan is normally required for UK contracts;
- Others for example programme, performance bonds etc.

Works information

The works information and site information (clauses 11.2(5) and (6)) together effectively form the equivalent of the traditional specification and drawings. The documents containing the works information provided by the *Employer* are identified in part one of the contract data. Any works information for the *Contractor's* design submitted by tenderers with their tenders is to be identified in part 2 of the contract data.

Site information

The documents in which the site information is contained are identified in part 1 of the contract data. Site information may include such issues as the subsoil investigation borehole records and any additional relevant information and reports.

16.8   Conclusion

This chapter contains a brief overview of some of the key provisions of the NEC3 contract which should be relevant to the construction cost manager. For a more in-depth understanding of the NEC3 contract readers are advised to refer to the *NEC Users' Group Newsletters*, the NEC3 *Guidance Notes* and Mitchell and Trebes' *Managing Reality* and if possible attend one of the excellent NEC conferences.

Using the NEC system requires a considerable commitment, both at the front end in defining the scope and throughout the project. However this high level of commitment and administrative support should ensure that there are no surprises at the end and secure an early final settlement.

The NEC system is based on the concept of cooperation with a proactive team-based approach. Great emphasis is placed on communications, programming and disciplined contract management by both parties. The NEC contract documents should be referred to throughout

---

Panel 16.2 Case study: The Eden Project, Cornwall, UK

The creation of the Eden project is an incredible story, the vision of one man – Tim Smit – and his challenge to create the eighth wonder of the world located in a disused clay pit in Cornwall. The Eden project's iconic biomes, the world's largest conservatories, are the symbol of a living theatre of plants and people, of regeneration and hope for the future.

The Eden Project was one of the largest projects funded by the National Lottery to celebrate the Millennium and involved a massive earthworks operation, construction of the two large dome structures, exhibition and catering facilities and horticultural planting.

The fundamental contractual requirement was to place a design and construct lump-sum contract with a guaranteed maximum price (GMP) with the successful contractor investing equity into the project. After 12 months of negotiations, the £60 million project was awarded to a joint venture of Sir Robert McAlpine and Alfred McAlpine, the first partnership between the two since they split in 1940. The design team, including the signature architect Nicholas Grimshaw & Partners, was then novated to the contractor. The contract was executed based on the NEC option C target cost – with GMP and savings/bonus provisions.

> One agreed a profit in advance with the contractor, obviating the need for them to find fault with others in order to prosper, and then the contract would be run on an open-book basis, meaning that all costs would be open to scrutiny from both sides. The scary thing was that for it to succeed everybody needed to work with the utmost good will; however, this would turn out to be one of the best decisions we ever made.
>
> (Tim Smit, 2002)

Sources: Smit, T. (2002); *Eden*, Corgi; No author (2000) 'Guaranteeing the Eden Experiment', *Building*, 24 March; Carter, T., *NEC Users' Group Newsletter*, Issue No. 16, p. 1; www.edenproject.com

the project; they should be considered manuals of project management as well as sets of conditions of contract.

The NEC contract poses fundamental challenges to the quantity surveying (QS) profession. As Martin Barnes critically observed reflecting on ten years of use of the NEC:

> But people tend to forget the amount of time which non-NEC contracts require to be committed to processing variations, claims and disputes. In the brave new world of NEC, you should retrain your quantity surveyors to become planners/estimators – using their ingenuity to come up with clever ways of building the thing instead of clever arguments over who should pay how much for work finished long ago.

(Barnes, 2003)

16.9   Question

1. Critically review the use of the NEC2 option C (target contract with activity schedule) under the NHS ProCure 21 scheme. Source: www.nhs-procure21.gov.uk (accessed 12 March 2007).

   Each year the ICE offers an annual set of examinations in Law and Contract Procedure (LCP); papers 2 and 3 contain questions on the NEC ECC contract. Questions with solutions can be found on the ICE website www.ice.org.uk/knowledge/contracts_examinations.asp (accessed 20 March 2007).

Bibliography

Barnes, M. (2003) 'NEC – a decade of success', *The NEC Users' Group Newsletter*, Issue No. 25, April

Bennett, J. and Baird, A. (2002) *NEC and Partnering: The Guide to Building Winning Teams*, Thomas Telford

Broome, J.C. (1997) 'Best practice with the New Engineering Contract', *Proc. Instn Civ. Engrs, Civ. Engng*, 120, May, 74–81

Broome, J.C. (1999) *The NEC Engineering and Construction Contract: A User's Guide*, Thomas Telford

Civil Engineering Surveyor, October 1997, November 1997, December 1997/January 1998, February 1998, March 1998, April 1998, May 1998, June 1998, July/August 1998, September 1998

Cox, A. and Thompson, I. (1996) 'Is the NEC going to succeed? – an examination of the Engineering and Construction Contract (alias the NEC 2nd edition)', *The International Construction Law Review*, pt 3, pp. 327–337

Eggleston, B. (2006) *The NEC 3 Engineering and Construction Contract: A Commentary*, 2nd edition, Blackwell Science

Forward, F. (2002) *The NEC Compared and Contrasted*, Thomas Telford

McInnis, J.A. (2001) *The New Engineering Contract: A Legal Commentary*, Thomas Telford

Mitchell, B. and Trebes, B. (2005) *Managing Reality* (set of 5 books: *Book 1 Introduction to the Engineering and Construction Contract*; *Book 2 Procuring an Engineering and Construction Contract*; *Book 3 Managing the Contract*; *Book 4 Managing Change*; *Book 5 Managing Procedures*), Thomas Telford

Perry, J.G. (1995) 'The New Engineering Contract: Principles of design and risk allocation', *Engineering, Construction and Architectural Management*, vol. 2, no. 3, pp. 197–208

The Institution of Civil Engineers (2005) *NEC3 Engineering and Construction Contract Guidance Notes ECC*, Thomas Telford

Valentine, D.G. (1996) 'The New Engineering Contract: Part 1 – a new language; part 2 – claims for extensions of time; part 3 – late completion and liquidated damages', *Construction Law Journal*, vol. 12, no. 5, pp. 305–332

Useful websites

www.ice.org/necuser
www.newengineeringcontract.com

17 FIDIC standard forms of international construction contract

17.1 Introduction

The FIDIC (the Fédération Internationale des Ingénieurs-Conseils) forms of contract, which were first published in 1957, are the most widely used standard international construction contracts. The FIDIC forms reflect the common standard for a wide range of international contracts around the world and are used with or without amendments or form the basis of bespoke public works contracts.

FIDIC, headquartered is in Switzerland, has a membership in more than 60 countries and represents most of the private consulting engineers in the world. FIDIC prepares and publishes a range of standard forms, which are updated on a regular basis after widespread consultation with its members, international contractors, major international banks including the World Bank and the International Bar Institution.

At first the FIDIC standard forms were intended for international use that is for projects where the client country was seeking participation of contractors from other countries. However in recent years the FIDIC forms have been increasingly used for domestic contracts where both client and contractor are of the same nationality. Following developments in the industry and after acknowledging anomalies in the old standard contracts, FIDIC produced in 1999 a new suite of forms to replace the existing forms.

17.2 The new forms

Late in 1999, FIDIC published four new standard forms; each of the new books for major works includes 'general conditions' together with guidance for the preparation of the 'particular conditions', a 'letter of tender', 'contract agreement' and 'adjudication agreements'.

Conditions of contract for construction: for building and engineering works designed by the employer (the new Red Book)

This is a traditional contract suitable where the 'employer' (or 'engineer') does most of the design. The 'contractor' constructs the works in accordance with the design provided by the employer – typically based on the drawings and specification but also including work specified in the engineer's instructions. However, the works may include some elements of contractor-designed civil, mechanical, electrical and/or construction works, for example, construction details and reinforcement.

Under the new *Red Book*, the engineer is required to administer the contract, monitor the construction works and certify payment; the employer is kept fully informed throughout the project and can make variations. Payment is based on a bill of quantities (BofQ) or lump sums for approved work done. This form effectively replaces and supersedes the FIDIC 4th-edition *Red Book* from 1987.

In 2005, FIDIC licensed the Multilateral Development Banks (MDB) to use the MDB *Harmonised Edition of the Construction Contract* for projects funded by the banks.

In using the FIDIC conditions it had been the regular practice of the MDBs to introduce additional clauses in the conditions of particular application or particular conditions in order to amend provisions contained in the FIDIC general conditions.

Furthermore the provisions in bid documents, including the additional clauses contained in the particular conditions, varied between MDBs. Hence, the need for the development of the MDB *Harmonized Form* which was first published in 2005 with an amended second version in 2006. The 2005 MDB *Harmonized Form* also includes sample forms for contract data, securities, bonds, guarantees and disputed board agreements.

FIDIC states not the intention to replace the standard 1999 contract – this is still available for use. However in reality the 2005 MDB *Harmonized Form* will surely become the standard contract throughout the developing world on projects funded by the international financing institutions.

Conditions of contract for plant and design-build: for electrical and mechanical plant and for building and engineering works designed by the contractor (the new Yellow Book)

Under this contract, the contractor (or supplier) is expected to do the majority of the design, not only of plant projects but also of various infrastructure and other types of work. The contractor's design is required to fulfil the employer's requirements, that is, an outline or performance specification prepared by the employer. Under this contract, the engineer administers the contract, monitors the manufacture and erection on site or construction work. The engineer also certifies payment, which are normally based on the achievement of milestones generally on a lump-sum basis.

Conditions of contract for EPC turnkey projects: engineer-procure-construct (the Silver Book)

This form is suitable on a PFI or PPP project where the concessionaire takes total responsibility for the financing, construction and operation of the project. Then the concessionaire (the 'employer') requires having an engineer-procure-construct (EPC) contract with the construction contractor where the contractor takes total responsibility for design and construction of the infrastructure or other facility.

Under this arrangement the employer does not expect to be involved in the day-to-day progress of the works. However, the employer expects a high degree of certainty that the agreed contract price and time will not be exceeded. Likewise, the contractor would expect to be paid a premium in return for bearing the extra risks involved.

This form can also be used on process-plant or power-plant projects where the employer provides the finance and wishes to implement the project on a fixed-price turnkey basis. However, in certain circumstances, the EPC *Yellow Book* is not suitable and the *Plant & D-B Book* is considered preferable.

Short form of contract for contracts of relatively small value (the Green Book)

This form is suitable for a relatively small contract, say under US$500,000, or the construction time is short, say less than six months, or the work involved is relatively simple or repetitive.

The new form is suitable for construction, electrical, mechanical or other engineering work with design by the employer (or his engineer/architect) or by the contractor.

The new *Red Book* is based on the traditional pattern of engineered-designed, contractor-built works. The new *Red Book* is similar to and an update of the old *Red Book*, but with some new features.

- Suitable for projects where main responsibility for design lies with employer (or his engineer);
- Some design may be carried out by the contractor;
- Administration of contract and supervision by engineer;
- Approval of work, payment etc. certified by engineer;
- Engineer acts as the impartial certifier and valuer;
- Engineer's decision as the first stage of dispute resolution (now renamed as decisions on matters of dissatisfaction);
- Work done is measured, payment according to BofQ;
- Option for payment on a lump-sum basis.

17.3　Balance of risk

All contracts are a compromise between the conflicting interests of the parties. The new *Red Book* attempts to allocate risks fairly between the parties. The basic principle is that the risk is allocated to the party that is best able to bear and control that risk. It follows therefore that the contractor can only be expected to be bound by and to price for conditions which are known or which he/she is able to foresee and reasonably price in his/her tender.

Typical risks carried by the contractor include: accuracy of estimate, appropriateness of method statement, costs of production, achieving estimated productivity, adherence to the programme, design and installation of temporary works, failure to obtain labour and materials, breakdown of construction equipment, failure by subcontractors, design of permanent works where undertaken by the contractor, weather, inflation etc.

17.4　Structure of the new *Red Book*

The new *Red Book* is divided into three sections.

1. The 'General Conditions' including the general conditions of dispute adjudication agreement and procedural rules;
2. Guidance for the preparation of particular conditions, including example forms of parent guarantee, tender security, performance security guarantee (both 'on demand' and 'upon default'), advance payment guarantee, retention money guarantee and payment guarantee by employer (Annexes A to G);
3. Forms of letter of tender, appendix to tender, contract agreement and dispute adjudication agreement (both for one-person and three-person Dispute Adjudication Board).

Contract documents

The contract documents as defined in sub-clause 1.1.1 [Definitions] are as follows:

- the contract agreement;
- the letter of acceptance;

- the letter of tender;
- the particular conditions;
- the general conditions;
- the specification;
- the drawings;
- the schedules (typically comprising a BofQ and a daywork schedule);
- any further documents listed in the contract agreement or in the letter of acceptance.

Sub-clause 1.5 identifies that the documents forming the contract are to be taken as mutually explanatory of one another. For the purpose of interpretation, the priority of the documents should be in accordance with the sequence listed above.

The parties

The parties to the contract are the employer and the contractor.

The employer plays very little part in the running of the contract. The contractor has to construct the works to the satisfaction of the engineer. The engineer is not a party to the contract, but is appointed by the employer and is given very wide powers to issue instructions to the contractor.

The engineer acts as an agent for the employer, for example, when he/she issues instructions under sub-clause 3.3 [Instructions of the Engineer] or under sub-clause 4.12 [Unforeseeable Physical Conditions]. The role of the engineer is thus not stated to be that of a wholly impartial intermediary unless such a role is specified in the particular conditions (FIDIC, 2000). However the engineer is required to act fairly between the parties. For example, under sub-clause 3.5 the engineer has to make a fair determination in accordance with the contract. Likewise, sub-clause 14.6 requires interim payment certificates to show the amount which the engineer fairly determines to be due.

Control of the project is undertaken at two levels.

- At the top, the engineer corresponds with the contractor;
- On site, the engineer is represented by the resident engineer (sub-clause 3.2), and the contractor by the contractor's representative (sub-clause 4.3).

17.5   The employer (clause 2)

The employer's role is limited to matters such as nominating the engineer; giving the contractor right of access to and possession of the site within the times stated in the appendix to tender; providing assistance to the contractor to obtain permits, licences and approvals; ensuring own personnel comply with safety and protection of the environment procedures; making payment upon certification; giving notice to the contractor of employer's claims and giving notice to terminate the contract. The employer gives all authority for the administration of the contract to the engineer.

Under sub-clause 2.4 following a request from the contractor the employer should within 28 days submit evidence that financial arrangements have been made which will allow the employer to pay the contractor punctually; failure to produce such evidence entitles the contractor to suspend or reduce the rate of work (sub-clause 16.1) and ultimately (if no evidence is received within 12 weeks) to terminate the contract. This sub-clause will be particularly relevant where the immediate client is a special-purpose vehicle (SPV) and is funded by loans.

Another new sub-clause 2.5 enables the employer if they consider themselves entitled either to any payment or an extension under the defects notification period under the contract to give notice to the contractor. The employer must follow the set of procedures if they consider themselves entitled to any payment and must give notice as soon as practicable and provide particulars of the claim.

17.6   The engineer (clause 3)

Engineer's duties and authority (sub-clause 3.1)

The engineer is not a party to the contract but essentially acts as an agent for the employer, with all their power to control the contractor described within the contract. The engineer is liable only to instruct the contractor provided one of the clauses in the contract give them that power. Most of the engineer's powers are derived from sub-clause 3.3 – Instructions of the Engineer.

In sub-clause 3.3 there is an obligation on the engineer to supply such additional or modified drawings which may be necessary for the execution of the works and the remedying of any defects.

Delegation by the engineer (sub-clause 3.2)

This sub-clause allows the engineer to assign duties and authority to assistants, that is, representatives permanently based on site. The assistants might include the resident engineer (RE) and/or independent inspectors. On larger projects there might be additional assistant resident engineers in order to support the RE. The authority of the engineer's assistants should be made in writing.

Note that the engineer should not delegate the authority to determine any matter in accordance with sub-clause 3.5 [Determinations].

The principal duties of the RE include the following:

- supervise and check that the works conforms to the drawings and specification;
- organize and supervise any tests;
- keep daily records: progress, labour and plant, problems, weather;
- examine the contractor's programme and method statement;
- check temporary works design;
- ensure that the site operations do not prejudice the safety of their own staff;
- measure the quantities of work;
- ensure that satisfactory records are kept (for payment);
- keep as built drawings.

Determinations (sub-clause 3.5)

Under this sub-clause the engineer, after consultation with the other parties, is required to agree and determine any matter. If agreement cannot be reached, the engineer should make a fair determination in accordance with the contract taking into account all the relevant factors. In practice, the QS(s) representing the employer and contractor will prepare much of the documentation forming the basis of the final negotiation.

17.7   The contractor (clause 4)

The contractor's general obligations (sub-clause 4.1)

The contractor is required to construct and complete the works (permanent and temporary), design the works (to the extent specified in contract), comply with engineer's instructions within the time stated and remedy any defects. Significantly, if the contract specifies that the contractor should design any part of the permanent works then this should be fit for such purposes (sub-clause 4.1(c)). It is noted that this obligation is more demanding than the employer's designers' obligation to design with reasonable skill and care. This could lead to some interesting disputes should difficulties arise as a result of any conflicts or anomalies between the employer's and contractor's designs.

The contractor is responsible for setting out safety and quality, providing the plant and contractor's documents specified in the contract, giving all notices, providing a bond or performance security and taking out the necessary insurances. Additionally, sub-clause 4.4 states that the contractor is wholly responsible for the performance of all subcontractors (including nominated).

Under sub-clause 1.12 the contractor is required to provide all such confidential information as the engineer might reasonably require to verify the contractor's compliance with the contract.

Site data (sub-clause 4.10)

Under sub-clause 4.10 the employer must make available to the contractor, 28 days prior to the submission of the tender, all relevant site data in their possession on the sub-surface and hydrological conditions including environmental aspects.

Sufficiency of the accepted contract amount (sub-clause 4.11)

Sub-clause 4.11 covers two closely related matters – the provision of information on the site by the employer and obligation of the contractor to inspect and examine the site to ensure the correctness and sufficiency of the 'accepted contract amount'.

The contractor is responsible for interpreting any information on the nature of the ground, subsoil and hydrological conditions, pipes and cables. Opinions, which may be included in such information, should not be relied upon. It is normally the case that soils reports are not part of the contract.

The contractor is only deemed to have satisfied himself as regards the nature of the ground, subsoil and hydrological conditions so far as practicable and reasonable, a contractor would not be expected to execute additional boreholes prior to tender. The contractor is deemed to have based his tender on the information made available to the employer and on his own inspection and examination.

Unforeseen ground conditions (sub-clause 4.12)

If the contractor encounters adverse physical conditions, which they consider have been unforeseeable, they should give notice to the engineer as soon as practicable [sub-clause 4.12 Unforeseeable Physical Conditions]. 'Unforeseeable' is defined by clause 1.1.6.8 as 'not being reasonably foreseeable by an experienced contractor by the date of the submission of the Tender'. The definition of physical conditions includes natural physical conditions and man-made and other physical obstructions and pollutants, including sub-surface and hydrological conditions but excluding climatic conditions.

Thus the employer is responsible for adverse physical conditions. Such conditions could be due to the following: lakes, high water table, geological faults, running sand etc.; services, mine workings, old structures etc. However such conditions should not be due to climatic conditions. Thus a flood after heavy rain would be a climatic condition but running sand probably not.

It is also necessary to ask if an experienced contractor could have reasonably foreseen the adverse condition. This question causes no end of arguments. If the engineer did not foresee and allow for conditions in his design, then it could be said that an experienced contractor could not reasonably foresee the conditions. The FIDIC *Contracts Guide* (FIDIC, 2000) gives the following guidance:

> For example, if the Time for Completion is three years, an experienced contractor might be expected to foresee an event which occurs (on average) once in every six years, but an event which occurs only once in every ten years might be regarded as Unforeseeable.

Under sub-clause 4.12 the contractor is required to give notice as soon as practicable of

1.   description of the physical condition;
2.   reasons why the contractor considers them to be unforeseeable.

If the contractor suffers delay and/or incurs cost due to these conditions, the contractor should be entitled subject to sub-clause 20.1 [Contractor's Claims] to

*   an extension of time for any delay, if completion would be delayed, under sub-clause 8.4 [Extension of Time for Completion];
*   payment of any such cost, which should be included in the contract price.

    (Note the definition of 'Cost' in sub-clause 1.1.43 includes overheads and similar charges but does not include profit.)

However there is a sting in the tail in this sub-clause in that the penultimate paragraph allows the engineer to reduce the cost if they consider the conditions more favourable than anticipated. Furthermore, a new concept allows the engineer to take account of any evidence of physical

---

Panel 17.1 Relevant legal case 1

The case *C.J. Pearce & Co Ltd v Hereford Corporation (1968)* 66 L.G.R. 647 concerned the construction of a sewer in a heading underneath a road crossing, under the terms of the 1955 ICE Conditions. The 'approximate line' of an ancient sewer was shown as diagonally crossing the line of the new sewer but its depth was not stated. However the witnesses for both parties accepted 'approximate', meant that the line of the old sewer might be 10 to 15 feet one side or other of the line shown.

Pearce, the contractor, fractured the old sewer, which was full of water resulting in flooding of the works. Following a meeting on site with the engineer, the contractor was required to cap off the old sewer at some distance from the site, sink a shaft at the far side of the road and work back using the open-cut method of excavation.

It was held that the contractor was bound to do the work and was not entitled to any additional expense. Even if a notice had been served under clause 12 it would have failed as the condition could have been 'reasonably foreseen'.

conditions foreseen by the contractor when submitting the tender, but the engineer is not bound by such evidence.

Encountering unforeseen physical obstructions is almost inevitable on major construction works. Even if the employer has instigated an extensive pre-contract site investigation, it is physically impossible to cover every square metre on the site with bore holes.

The history of tunnelling is littered with claims for 'unforeseeable physical conditions'. Tunnelling contractors develop their method statement and order their construction equipment based on the ground condition information given at tender stage. If rock is anticipated, then drilling machines or a drill and blast approach may be appropriate. If there is a high water table or poor ground conditions, a tunnelling shield working in compressed air may be chosen. These major items of construction equipment are usually purchased specifically for the job at hand and may take anything up to six months to manufacture.

If the contractor encounters different ground conditions than those anticipated, this could have major repercussions on the costs and the completion date. On the Carsington Dam tunnelling project for Severn Trent Water in the late 1970s Mowlem ordered four large drilling machines. In the event, the ground conditions proved worse than anticipated and the drilling rigs were discarded as being unsuitable. A tunnelling shield using compressed air was introduced, but progress was slow due to isolated pockets of rock. Three years into the Carsington project the engineer instructed additional site investigation! Indeed the only tunnels that went as planned on this large project were the hand-dug tunnels. There is a saying in the world of tunnelling that the only accurate ground-investigation survey is the tunnel itself!

On the first stage of the Hong Kong MTR (HKMTRC) The Gammon Kier Lilley JV had planned to construct the external perimeter walls to the station boxes in North Natham Road using large-diameter piling equipment specially manufactured for the project in the UK. In order to accommodate these massive piling rigs it was necessary first to remove all the overhanging canopies in the tightly congested high street. In the event, isolated granite boulders were encountered which meant that the expensive equipment could not perform the work as planned.

The contractor developed a solution for the outer wall construction involving a combination of grout curtains around the outer box walls and hand-dug caissons based on a tried and tested approach which had been used in China and elsewhere for centuries. The substantial additional costs became the subject of a major claim under the unforeseeable physical conditions stipulated in sub-clause 4.12.

Progress reports (sub-clause 4.21)

If required by the contract, the contractor should provide regular progress reports. The regular and timely production of this detailed list of documents is critical. Not only do they act as a monitor on progress they also act as a condition of payment. Under sub-clause 14.3 payment will be made only within 28 days of receipt of the application for payment and the supporting documents. This report will be a substantial document and must be submitted within seven days from the last day of the relevant month, which means five working days (Totterill, 2006).

17.8   Commencement, delays and suspension (clause 8)

Commencement of works (sub-clause 8.1)

The engineer should give the contractor not less than seven days notice of the commencement date. Unless otherwise stated in the particular 'Conditions', the commencement date should be within 42 days after the contractor receives the letter of acceptance.

Time for completion (sub-clause 8.2)

The contractor should complete the whole of the works within the time for completion for the works.

The programme (sub-clause 8.3)

The contractor makes assumptions based on the borehole logs and soil-investigation reports. They plan the mode, manner and methods of working; design the temporary works and estimate the subsequent productivity rates based on the information given to them at tender stage. This information thus forms the basis for their tender as well as their programme. If the contractor does extra work they are entitled to reasonable payment.

A project is required to be programmed many times, with different levels of detail, throughout the project cycle. One of the most important means by which the contractor can increase their profit is by finishing a project early, thus saving on-site overheads.

Within 28 days after receiving the notice under sub-clause 8.1 [Commencement of Works] the contractor should submit to the engineer a detailed time programme. The contractor should also submit a revised programme when the previous programme is inconsistent with actual progress or the contractor's obligations. Each programme should include

- the order in which the contractor intends to carry out the works, including the anticipated timing of each stage of design (if any), contractor's documents, procurement, manufacture of plant, delivery to site, construction, erection and testing;
- each of these stages for work by each nominated subcontractor (as defined in clause 5 [Nominated Subcontractors]);
- the sequence and timing of inspections and tests specified in the contract, and a supporting report which includes: general description of the methods which the contractor intends to adopt, and of major stages, in the execution of the works, and details showing the contractor's reasonable estimate of the number of each class of contractor's personnel and of each type of contractor's equipment, required on the site for each major stage.

   Points to note:

- the programme is not a contract document;
- the programme is not for approval of the engineer;
- form of the programme is not dictated; however, some employers and engineers might wish to specify the format, for example, Asta Powerproject or Primavera.

Note that the engineer has 21 days from receipt of the programme to check and ensure that it complies with the contract. If at any time the engineer gives notice to the contractor that a programme fails to comply with the contract or to be consistent with actual progress and the contractor's stated intentions, the contractor should submit a revised programme.

The author recollects that while working for the HKMTRC, a contractor submitted a grossly over-optimistic programme in terms of productivity. No approval was given for this programme; the employer, through the engineer, prepared a realistic version of the programme which was later used for the validation of claims for extension of time.

Under sub-clause 8.3 the contractor is required to give advance notice or early warning to the engineer of probable future events or circumstances which may adversely affect the works. This requirement indicates a desire on behalf of the employer to instil a proactive team-based philosophy to solving problems.

The engineer requires the construction programme to

1. organize their office regarding drawings, schedules, nominated subcontractors;
2. organize their engineering staff and inspectors;
3. monitor the contractor's progress;
4. establish the cash-flow (for the employer);
5. monitor claims for extension of time.

The contractor requires a construction programme to

1. organize subcontractors, procure construction equipment, labour and materials;
2. plan and control the works;
3. establish the basis of claims (extension of time);
4. establish the cash flow.

Contractors often develop strategies in order to support delay claims by various means: by detailing early requirements for all further design information; by showing all items as critical on the network (with no float); by including an early *date for completion*. However in the case of *Glenlion Construction Ltd* v *The Guinness Trust (1987)* 39 BLR 89 it was held that an architect was under no obligation to provide the contractor with information early even though the contractor's programme showed them completing in 101 weeks in contrast to the 114 weeks in the contract.

Some employers (and/or their engineers) require tenderers to include details of their methods and their intended programme with their tender. The reasons for this include

*   obtaining information to be used in the appraisal of tenders;
*   obtaining information to assist the employer and/or engineer in fulfilment of their own obligations;

Panel 17.2 Relevant legal case 2

The case of *Yorkshire Water Authority* v *Sir Alfred McAlpine Ltd (1985)* 32 BLR 114 concerned a dispute on the £7 million outlet tunnel at Grimworth Reservoir in North Yorkshire under the ICE 5th. The contractors were required to submit with their tender a programme in bar chart format together with a method statement showing that they had taken note of certain specified phasing requirements – in particular that the upstream work preceded the downstream requirements.

In the event, this proved impossible and the contractor proceeded with the downstream work and sought a variation under clause 51(1) ' . . . such changes may include . . . changes in the specified sequence method or timing of construction.'

The court held that

*   the incorporation of the method statement into the contract imposed an obligation on the contractor to follow it so far as was legally or physically possible;
*   the method statement, therefore, became a specified method of construction and the contractor was entitled to a variation order and payment accordingly.

Mr Justice Skinner commented, 'The plaintiff [Yorkshire Water] could have kept the programme and methods as the sole responsibility of the contractor under clause 14(1) and (3) – the risks would then have been the respondent's [McAlpine] throughout.'

- obtaining information to assist the employer and/or engineer in coordinating the contract works with other activities;
- obtaining information in the hope of controlling the contractor's operations and their scope for claims.

Extensions of time for completion (sub-clause 8.4)

The contractor is under strict duty to complete on time except to the extent that they are prevented from doing so by the employer or is given relief by the express provisions of the contract. The effect of extending time is to maintain the contractor's obligation to complete within a defined time and failure by the contractor to do so leaves them liable to damages, either liquidated or general according to the terms of the contract (Egglestone, 2001).

Sub-clause 8.4 defines the events, which entitle the contractor to an extension of time and sets out the procedures and rules for the contractor and engineer to follow.

Extension of time clauses have various purposes

1. to retain a defined time for completion;
2. to preserve the employer's right to liquidated damages against acts of prevention;
3. to give the contractor relief from their strict duty to complete on time in respect of delays caused by designated neutral events.

Under sub-clause 8.4 items giving grounds for an extension of time include the following:

- a variation (unless an adjustment to the 'time for completion' has been agreed under sub-clause 13.3 [Variation Procedure]) or other substantial change in quantity of an item of work included in contract;
- a cause of delay giving an entitlement to extension of time under a sub-clause of these conditions;
- exceptionally adverse climatic conditions;
- unforeseeable shortage in the availability of personnel or goods caused by epidemic or governmental actions, or any delay, impediment or prevention caused by or attributable to the employer, the employer's personnel or the employer's other contractors on the site.

If the contractor considers themselves to be entitled to an extension of time for completion, then they should give notice to the engineer in accordance with sub-clause 20.1 [Contractor's Claims].

Of itself, an extension of time under the new *Red Book* gives no entitlement to payment to the contractor and the question of whether a particular delay is reimbursable or non-reimbursable is properly determined from the cause of delay (and proof of cost arising) rather than whether or not an extension of time has been granted (Egglestone, 2001). If the contractor also seeks reimbursement of additional costs and reasonable profit, then the claim has to be made under that particular sub-clause.

The following clauses carry an entitlement to an extension of time plus cost and a reasonable profit:

1.9   Delayed drawings and instructions
2.1   Right of access to site
4.7   Setting out

7.4   Testing
10.3   Interference with tests on completion
16.1   Contractor's entitlement to suspend work.

The following sub-clauses will permit an entitlement to an extension of time plus cost only in certain situations:

4.12   Unforeseeable physical condition
4.24   Fossils
8.9   Suspension initiated by the employer
13.7   Adjustment for changes in Legislation
17.4   Consequences of employer's risks
19.4   Consequences of Force Majeure.

The following sub-clauses will permit an entitlement to an extension of time only.

Delays caused by authorities

Rate of progress (sub-clause 8.6)

If the actual rate of progress is too slow to complete within the time for completion and/or progress has fallen, or will fall, behind the current programme, the engineer can require the contractor to produce a revised programme and revised working methods in order to expedite progress and complete within the time for completion. The engineer can do this only if the reason for the delay is other than a cause listed in the sub-clause 8.4 [Extension of time for completion]. Unless the engineer objects the contractor should adopt the revised working methods at their own risk and cost.

Delay damages (sub-clause 8.7)

If the contractor fails to complete within the time for completion calculated from the commencement date, they should pay to the employer a sum (expressed per day or per week) which represents a genuine pre-estimate of the damages likely to be suffered (not a penalty). These delay damages should be the sum stated in the Appendix to Tender; it is noted that the FIDIC Guide recommends a limit on international projects generally varying between 5% and 15%.

17.9   Measurement and evaluation (clause 12)

Works to be measured (sub-clause 12.1)

The engineer is responsible for measuring the works; they are required to give notice to the contractor's representative to assist in making the measurement and supplying any particulars. If the contractor fails to attend the measurement made by or on behalf of the engineer, the measurement is accepted as accurate. It is noted that the contractor has only 14 days to give notice of any disagreement.

The new *Red Book* is essentially a measure and value contract with the works being subject to re-measurement upon completion. The measurement is normally calculated using the latest approved engineer's drawings valued based on the rates in the BofQ(s) or schedules. However some measured items will need to measured in situ on site based upon agreed records, for example excavation in rock.

It is noted that if a lump-sum contract is to be adopted the particular conditions recommend that amendments are made to the wording of clauses 12, 13 and 14.

Method of measurement (sub-clause 12.2)

Except as otherwise stated, measurement should be based on the net quantity of each item of the permanent works. The method of measurement should be in accordance with the BofQ or other schedules. It is important that the rules of measurement are clearly defined in the contract.

The case studies below indicate two different methods of measurement, one based on partial re-measurement in accordance with the terms defined in the schedules and the other on full re-measurement based on a specially written standard method of measurement.

Evaluation (sub-clause 12.3)

Under this sub-clause the engineer is required to agree or determine the contract price by evaluating each item of work, applying the measurement agreed and the appropriate rate or price for the item.

Sub-clause 12.3 indicates that the appropriate rate should be the rate or price specified in the contract. However, a new rate or price should be appropriate for an item of work if

(a) (i)   the measured quantity of the item is changed by more than 10% from the quantity of this item in the BofQ or any other schedule;

Panel 17.3 Case studies: typical measurement provisions within FIDIC contracts

Case study 1: Teesside Jetties Marine Terminal for US client Phillips Petroleum

The tender documents included a detailed specification and drawings of the permanent works. An activity schedule was included which listed the major components in the project; bidding contractors were required to insert lump-sum prices against the activities described.

Two elements of the work were further analysed and measured in detail in a mini-bill of quantities – the 36" diameter steel tube piles together with the integral sea-bed anchorages. These items were selected for re-measurement due to the uncertain nature of the ground conditions and anticipated varying lengths required. The piling and anchorages were re-measured on completion based on the agreed driven records and priced at the tender rates where appropriate. New rates were negotiated for pile extensions.

Case study 2: HKMTRC – stage 3 (Island Line)

The project was designed in detail by the client's design consultants. Comprehensive bills of quantities covering all the permanent work designed were prepared. Some critical temporary works, which were designed by the client's designers, for example, ground-water control systems, were measured and included in the bills of quantities.

The bills of quantities were prepared based on the MTRC's own method of measurement; this was similar to CESSSM2 but had expanded sections covering tunnelling and architectural finishes. All work was subject to re-measurement on completion based on the permanent works, finalized design drawings or records agreed on site, for example, records were produced for every linear metre of tunnel excavation.

(ii)   this change in quantity multiplied by such specified rate for this item exceeds 0.01% of the accepted contract amount;

(iii)   this change in quantity directly changes the cost per unit quantity of this item by more than 1%;

(iv)   this item is not specified in the contract as a 'fixed rate item'.

or

(b)   (i)   the work is instructed under clause 13 [variations and adjustments];

(ii)   no rate or price is specified in the contract for this item;

(iii)   no specified rate is appropriate because the item of work is not of similar character, or is not executed under similar conditions, as any item in the contract.

The new *Red Book* identifies each new rate or price should be derived from any relevant rates or prices in the contract, with reasonable adjustments to take account of matters described in sub-paragraph (a) and/or (b). If no rates are or prices are relevant for the derivation of a new rate it should be derived from the reasonable cost of executing the work, together with a reasonable profit, taking account of other relevant matters.

Contractors frequently submit accounts for varied work in the form of daywork sheets. Such a valuation frequently suits contractors as reimbursement is essentially made on a cost-plus-profit basis with no incentive for efficient working. However this basis is only formally applicable where the engineer orders that a variation be executed on a daywork basis (sub-clause 13.6).

There are several general principles when negotiating the value of variations. Try to follow the same principles that the contractor used in calculating his/her rates for the tender. Establish a valuation fair to both parties, that is, cost plus a reasonable percentage for profit, with a deduction of any proven inefficiency by the contractor. The market rate should be taken into consideration or used completely. However only in exceptional cases should the basis of the valuation from BofQ rates be abandoned.

Omissions (sub-clause 12.4)

In essence, this sub-clause entitles the contractor to compensation for costs reasonably incurred in the expectation of carrying out work subsequently omitted under the variation.

17.10   Variations and adjustments (clause 13)

Right to vary (sub-clause 13.1)

Variations may be initiated by the engineer at any time prior to issuing the 'taking-over certificate for the works', either by an instruction or by a request for the contractor to submit a proposal. Variations might include

1.   changes to the quantities of any item of work included in the contract (however such changes do not necessarily constitute a variation);
2.   changes to the quality and other characteristics of any item of work;
3.   changes to the levels, positions and/or dimensions of any part of the works;
4.   omission of any work, unless it be carried out by others;
5.   any additional work, plant, materials or services necessary for the permanent works, including any associated Tests on Completion, boreholes and other testing and exploratory work;
6.   changes to the sequence or timing of the execution of the works.

Note: The engineer cannot order changes to the contract itself; wherever practicable all variations should be in writing; changes in quantity (where correction of error in BofQ) do not need a variation order – for oral instructions see sub-clause 3.3.

Value engineering (sub-clause 13.2)

Value management (VM) and value engineering (VE) are important project management concepts which encourage the parties to innovate and seek best-value solutions throughout design and construction phases of a project.

Under sub-clause 13.2 the contractor is encouraged to submit proposals which may accelerate completion, reduce the cost to the employer, improve the efficiency or value to the employer, or otherwise be of benefit to the employer. If the proposal is accepted, the contractor may claim half of the saving in contract value.

Panel 17.4 Case study: LimeHouse Link Road Tunnel, London Docklands

In order to secure maximum benefit, value engineering is best carried out at an early stage in the design process. However there are many examples of value engineering undertaken on major projects during the construction process. For example, on London Dockland's £250 million Limehouse Link highway tunnel project, innovative thinking and value engineering were used to eliminate the substantial temporary steel strutting system for the diaphragm walls on the major cut and cover project.

Work started on site in November 1989 but soon encountered problems that caused delay and increased costs. A variation agreement between the London Docklands Development Corporation and the contractor Balfour Beatty-Amec was subsequently negotiated. This included the addition of a value engineering clause to the contract in March 1991. The value engineering clause facilitated the introduction of the observational method and created opportunities to introduce design changes that increased the speed of construction and substantially decreased cost. Operational safety was also enhanced. The principal need was to reduce delay to the programme. (Powderham, 2002, www.tunnels.mottmac.com/files/page/1614/OM-LearningProjects.pdf [accessed 19 February 2007]).

Variation procedure (sub-clause 13.3)

If the engineer requests a proposal, prior to instructing a variation, the contractor should respond in writing as soon as practicable, either by giving the reason why they could not comply or by submitting

1. a description of the proposed work and a programme for its execution;
2. the contractor's proposal for any necessary modifications to the programme and to the time for completion;
3. the contractor's proposal for evaluation of the variation.

The engineer should, as soon as practicable, respond with approval, disapproval or comments.

Variations should be evaluated in accordance with clause 12 [Measurement and Evaluation], unless the engineer instructs or approves otherwise in accordance with this clause.

Traditionally variations have been valued after the works are completed with initial valuation based on the rates contained in the (BofQ) or pro-rata thereto. These rates are often considered inappropriate and the variation is valued on a 'fair and reasonable basis'.

The FIDIC new *Red Book* introduces the concept of requiring the contractor to submit a proposal before the varied work is instructed. This is a fundamental change in philosophy for if the value of the variation and associated programme implications are accepted by the engineer prior to instruction the risk is shifted from the employer to contractor. The contractor thus has the incentive to work efficiently on the varied work.

The author recollects on a project based on the FIDIC form the contractor's representative being asked by the engineer to give a quotation for a major variation. The initial lump-sum quotation of £300,000 was rejected, likewise a more detailed quotation in the sum of £400,000. Eventually the contractor was instructed to carry out the work on dayworks basis – final sum £500,000! From the employer's viewpoint settling early often means settling cheaply. It is easy for contractors to underestimate the anticipated final cost and time implication of a variation, particularly if subcontractors are involved.

Payment in applicable currencies (sub-clause 13.4)

This sub-clause permits reimbursement in more than one currency, if appropriate for a variation under sub-clause 13. The contract currency proportions may or may not be appropriate to the variation; however, they should still be taken into account in the evaluation of the variation.

The author recollects that the contractor was required to include for three currencies in the tender bid on the Teesside Jetties project for Phillips Petroleum. Under this FIDIC Contract the bid included: UK Sterling for the main work, German Deutschmarks (which covered the purchase of 24"×54" large diameter piles from Mannesmann) and Dutch Guilders (which covered the purchase of the oil-loading arms from OWECo in Holland).

Provisional sums (sub-clause 13.5)

Each provisional sum should be used only in whole or part in accordance with the engineer's instructions. Provisional sums are often included in the BofQ for parts of the works which are not required to be priced at the contractor's risk, for example, for uncertain works, removal of contaminated ground etc.

The provisional sum can be executed by the contractor and valued under clause 13.3 [variation procedure] or by a nominated subcontractor, as defined in clause 5 [nominated subcontractors] and valued on a cost-plus basis.

Daywork (sub-clause 13.6)

For work of a minor or incidental nature, the engineer may instruct that a variation be executed on a daywork basis to be valued in accordance with the daywork schedule included in the contract.

Daywork is usually instructed for work which cannot easily be measured, for example, search for and locate services – if location not as per drawings. The labour, materials and construction equipment involved in the work should be recorded on a daily basis and signed by the engineer. A priced summary of the dayworks should be submitted to the engineer prior to their inclusion each month in the next statement under sub-clause 14.3 [application for interim payment certificates].

It is important that the daywork record sheets are accurately recorded by the site engineer and accurately priced by the QS. The site engineers should be familiar with the daywork schedule as included in the contract and should ensure that all reimbursable items are included on the agreed record sheets. Likewise the contractor's QS should ensure that the rates are calculated accurately in accordance with the definitions in the contract. It is noted that E&F Spon's *Civil Engineering and Highway Works* (updated each year) identifies a number of items which might not have been included on the daily-record sheets and recommends the addition of appropriate percentages to cover these items.

- General servicing of plant
- Fuel distribution
- Welfare facilities
- Handling and offloading materials.

If the engineer/engineer's assistants cannot agree that the work is additional, then the records should be checked on a daily basis and signed 'Agreed for record purposes only'.

In practice, the dayworks schedule in the contract may not be fully comprehensive and include all the types of labour and construction equipment on site. The following documents, which are used under the ICE Conditions in the UK, might prove useful in the negotiation process:

- The Civil Engineering Contractors' Association, *Schedules of Dayworks Carried out Incidental to Contract Work* (latest version).
- *The Reference Manual for Construction Plant (incorporating The Surveyors' Guide to Civil Engineering Plant)* (published by the Institution of Civil Engineering Surveyors).

In the past, daywork records have been used by contractors on civil engineering projects to justify the amount due under claims. This approach should be considered only as last resort and the employer's representatives might seek discounts on the rates quoted depending on the circumstances; an example would be the hourly rates for contractor-owned construction equipment, which might not be appropriate on long-term hire.

Adjustment for changes in legislation (sub-clause 13.7)

This sub-clause protects the parties from the consequences of changes in legislation in the country made after the base date (28 days prior to the latest date for submission of the tender). It is noted that this sub-clause also allows the employer to reduce the contract price under sub-clause 2.5 [employer's claims]. Therefore both the contractor's and employer's cost managers should be alert to changes in legislation.

Adjustment for changes in cost (sub-clause 13.8)

Under a fixed-price contract no adjustments are made for escalation of costs. The introduction of this sub-clause provides a formula to adjust the contract value to reflect increased costs due to inflation.

The formulae includes the following components: non-adjustment element; coefficients representing the estimated proportion of each cost element such as labour, equipment and materials; the current monthly cost indices or reference prices for the period; base-cost indices or reference prices.

This sub-clause is appropriate for complex long-term contracts and/or where there is high inflation in the country concerned. The indices and formulae might also be appropriate in establishing the amount due for additional payment under sub-clause 20.1 [contractor's claims] where there has been severe delay outside the control of the contractor.

17.11   Contract price and payments (clause 14)

The contract price (sub-clause 14.1)

Unless otherwise stated in the particular conditions

1. the contract price should be agreed or determined under sub-clause 12.3 [evaluation] and subject to adjustments in accordance with the contract;
2. the contractor should pay all taxes, duties and fees; the contract price should be adjusted only for costs valued under sub-clause 13.7 [adjustments for changes in legislation];
3. any quantities set out in the BofQ or other schedule are the estimated quantities and are not to be taken as the actual or correct quantities;
4. the contractor should submit to the engineer within 28 days after the commencement date, a proposed breakdown of each lump-sum price in the schedules. The engineer might take account of the breakdown when preparing payment certificates, but is not bound by them.

Advance payment (sub-clause 14.2)

In accordance with this sub-clause the employer should make an advance payment, as an interest-free loan for mobilization, when the contractor submits a guarantee. The number and timing of instalments and the applicable currencies should be stated in the Appendix to Tender. Thus the right to advance payment for mobilization is not automatic.

The engineer should issue an interim payment certificate for the first instalment after receiving a statement under sub-clause 14.3 [application for interim payment certificates] and after the employer receives (1) the performance security in accordance with sub-clause 4.2 [performance security] and (ii) a guarantee in amounts and currencies equal to the advance payment.

The advance payment is repaid through percentage deductions in the payment certificates, which commence when the total of certified interim payments exceeds 10% of the accepted contract amount minus any provisional sums.

This sub-clause enables the contractor to recover the substantial additional costs involved in mobilizing international projects. These costs can include staffing costs including flights and accommodation, rented offices and site accommodation; fabrication yards and workshops; manufacture and shipping of construction equipment including specialist earthmoving equipment and dump trucks, cranage, piling equipment, tunnelling shields, rock drills, grouting equipment etc.

Application for interim payment certificates (sub-clause 14.3)

This sub-clause requires the contractor to submit a statement in six copies to the engineer at the end of each month, showing in details the amounts to which the contractor considers him/herself entitled together with supporting documents which includes the report on progress during the last month in accordance with sub-clause 4.21 [progress reports].

Sub-clause 14.3 states that '[t]he Statement shall include the following items, expressed in the various currencies in which the Contract Price is payable, in the sequence listed':

(a) the estimated contract value of the Works and the Contractor's Documents produced up to the end of the month (including variations but excluding items described in sub-paragraphs (b) to (g) below);

(b) any amounts to be added and deducted for changes in legislation and changes in cost, in accordance with Sub-Clause 13.7 [Adjustment for Changes in Legislation] and Sub-Clause 13.8 [Adjustment for Changes in Cost];

(c) any amount to be deducted for retention, calculated by applying the percentage of retention stated in the Appendix to Tender to the total of the above amounts, until the amount so retained by the Employer reaches the limit of Retention Money (if any) stated in the Appendix to Tender;

(d) any amounts to be added or deducted for the advance payment and repayments in accordance with Sub-Clause 14.2 [Advance Payment];

(e) any amounts to be added and deducted for Plant and Materials in accordance with Sub-Clause 14.5 [Plant and Materials Intended for the Works];

(f) any other additions or deductions which may have become due under the Contract or otherwise, including those under Clause 20 [Claims, Disputes, and Arbitration]; and;

(g) the deduction of amounts certified in all previous Payment Certificates.

The engineer considers this statement and supporting documents and within 28 days after receiving the statement, after making any correction or modification deemed appropriate and issues to the employer an interim payment certificate showing the amount due to the Contractor.

Schedule of payments (sub-clause 14.4)

This sub-clause takes effect when the contract states the interim payments are to be made in accordance with a specific schedule of payments. The sub-clause allows the engineer to agree or determine revised instalments if the progress is less than that on which the instalments were based.

If the contract does not include a schedule of payments the final paragraph requires the contractor to submit non-binding estimates of the payment amount it anticipates becoming due during the each quarterly period.

Plant and materials intended for the works (sub-clause 14.5)

If this sub-clause applies interim payment certificates should include an amount for plant and materials which have been sent to site for incorporation in the permanent works. Note that only those items which are listed on the Appendix to Tender are eligible for inclusion. The contractor is required to keep satisfactory records and submit a statement of cost supported by evidence. Payment will be based on 80% of the engineer's determination of the cost. As with sub-clause 14.2 [advance payment] a guarantee must be provided by the contractor if an advance payment is to be made.

Issue of interim payment certificates (sub-clause 14.6)

This sub-clause requires the engineer to issue to the employer an interim payment certificate stating the amount which the engineer fairly determines to be due. This certificate should be issued to employer within 28 days after receiving the contractor's statement and supporting documents.

However, it is noted that the interim payment certificate can be withheld if the employer has not received the performance security or if the amount to be certified is less than the minimum amount stated in the appendix to the tender. Furthermore the engineer can make deductions if the work is not carried out in accordance with the contract or if the contractor is failing to perform any obligation in accordance with the contract. It is clear that the engineer cannot withhold the issue of an interim payment certificate because the progress report had not been submitted (Totterill, 2006).

Payment (sub-clause 14.7)

This important sub-clause sets out the dates when the employer must pay the contractor

1. the first instalment of the advance payment within 42 days after issuing the Letter of Acceptance or within 21 days after receiving documents in accordance with sub-clause 4.2 [performance security] and sub-clause 14.2 [advance payment], whichever is later.
2. the amount certified in each interim payment certificate within 56 days after the engineer receives the statement and supporting documentation;
3. the amount certified in the final payment certificate within 56 days after the employer receives this payment certificate.

Delayed payment (sub-clause 14.8)

If payment from the employer is received late this sub-clause 14.7 entitles the contractor the right to automatically claim finance charges calculated at a rate of 3% above the base rate.

Payment of retention money (sub-clause 14.9)

This sub-clause identifies the dates for the release of the retention monies.

- The first half is released after the issue of the taking-over certificate;
- If the taking-over certificate is for part of works then the retention should be calculated based on 40% of the value of the section;
- The outstanding balance is released after the expiry of the defects correction period.

However if any work remains to be executed under sub-clause 11 [defects liability] the engineer would be entitled to withhold certification of the estimated cost of the work.

The contractor's QS should ensure that the application for retention release is made in the next application following the receipt of the take-over certificate.

Statement at completion (sub-clause 14.10)

This sub-clause requires the contractor to submit to the engineer six copies of its completion statement within 84 days after receiving the taking-over certificate. The completion certificate should show: the value of all work done in accordance with the contract; any further sums which the contractor considers himself entitled and an estimate of any other amounts.

Application for final payment certificate (sub-clause 14.11)

This sub-clause requires the contractor to submit six copies of the draft final statement to the engineer within 56 days after receiving the performance certificate. The draft final statement should show the value of all work done and the value of all further sums which the contractor considers due.

17.12   Claims, dispute and arbitration (clause 20)

Contractor's claims (sub-clause 20.1)

If the contractor considers themselves entitled to any extension of time or for additional payment, under any clause in the conditions or otherwise in connection with the contract, they should give notice to the engineer, describing the circumstances. The notice should be issued as soon as practicable and not later than 28 days after the contractor becomes aware of the circumstances (sub-clause 20.1). It is essential that the contractor gives notice within the 28-day period as failure to give such notice will result in the employer being discharged from liability.

The contractor should keep contemporary records to substantiate any claim. Without admitting liability the engineer may after receiving any notice of claim monitor the record keeping. The new *Red Book* also requires the contractor within 42 days to send to the engineer a fully detailed claim which includes full supporting particulars.

Facts could be based on measurements, labour and plant time sheets, daily diaries, reports, photographs, site minutes and correspondence.

Within 42 days after receiving the claim the engineer should respond with approval, or disapproval and detailed comments. He may also request further particulars but should respond on the principles of the claim within the timescale.

Each payment certificate should include when such amounts for any claim, which have been reasonably substantiated should be included in the payment certificate. Updated claims should be submitted on a monthly basis with the final claim submitted within 28 days of the end of the event unless otherwise agreed.

Appointment of the Dispute Adjudication Board (sub-clause 20.2)

The Dispute Adjudication Board (DAB) is a feature intended to improve harmony between the contractor and employer by resolving disputes promptly (Roe, 2007). The DAB comprises one or three members who are suitably qualified. The DAB procedure is the primary method for resolving disputes under the contract. It replaces the process of decision-making by the engineer and must occur prior to any reference to arbitration.

It is to be noted that the DAB decision is binding on both parties, unless revised by amicable settlement or an arbitral award (sub-clause 20.4). However in practice, most funding agencies prefer that claims are resolved in accordance with sub-clause 20.5 [amicable settlement] (Papworth, nd).

17.13   Conclusion

This chapter has summarized briefly some of the key provisions under the new *Red Book* where the employer (or engineer) does most of the design and the contractor constructs the works. The engineer has a key role to play under these conditions in administration of the contract, monitor the construction works and certify payments.

This traditional form of construction contract forms the basis for contract conditions on many major projects executed throughout the world. The QS, either representing the client or contractor, has a key role to play under this form of contract and is a valuable member of the commercial team.

Important pre-contract duties could include preparation of the tender documents, schedules or detailed BofQ and assisting with tender evaluation. Post-contract duties could include re-measurement, preparing interim valuations, valuing variations, assisting with VE submissions, assisting with claims evaluation, settling the final account and financial reporting.

17.14   Questions

1. Identify five risks carried by the employer under the new *Red Book* and state how these might be avoided or minimized.
2. Identify the sub-clauses within the contract which provide the employer with the right to claim from the contractor.
3. Identify the items for consideration by the engineer after receiving a notice of unforeseeable physical conditions from the contractor.
4. Compare and contrast how FIDIC new *Red Book* and the NEC3 *Contract* deals with risks for unforeseen site conditions.

Solution: Refer to Ndekugri and McDonnell's (1999) article which compared the provisions in FIDIC 1987 with those in the NEC 2nd edition.

5. Identify those items which should be included in the progress report and state who should provide the information to the contractor and when.
6. Identify the main contractor's site overheads on a typical civil engineering project.
7. Summarize the procedures with which the contractor must comply in order to convert an oral instruction received from the engineer/delegated assistant into a valid engineer's instruction.
8. Identify how each of the items (a) to (g) in sub-clause 14.3 will be evaluated each month.
9. Draw a diagram which clearly shows the typical sequence of payment events envisaged in clause 14.
10. Identify the innovations in the new *Red Book* which (a) favour the contractor, and (b) favour the employer. For solution, see Seppala (2001).

Acknowledgements

The author is deeply grateful to FIDIC for allowing the clauses in the Conditions of Contract to be reproduced and for guidance from the FIDIC Contracts Guide. Copies of the Conditions of Contract and other FIDIC publications can be obtained from the FIDIC bookshop, Box 311 CH-1215 Geneva 15, Switzerland (tel: +41–22–799–4900; fax: +41–22–799–4901; *fidic. pub@pobox.com*; www.fidic.org/bookshop).

Bibliography

Abrahamson, M.W. (1985) *Engineering Law and the I.C.E. Contracts*, 4th edition, Elsevier Applied Science
Eggleston, B. (2001) *The ICE Conditions of Contract*, 7th edition, Blackwell Scientific
European International Contractors (2003) *EIC Contractor's Guide to the FIDIC Conditions of Contract for Construction*, EIC

FIDIC (2000) *FIDIC Contracts Guide to the Construction, Plant and Design-Build and EPC/Turnkey Contracts*, 1st edition, FIDIC

Ndekugri, I. and McDonnell, B. (1999) 'Differing site conditions risks: a FIDIC/engineering and construction contract comparison', *Engineering Construction and Architectural Management*, vol. 6, no. 2, pp. 177–187

Papworth, J. (no date) 'Claims under the New FIDIC Conditions of Contract' available online at: www.awestcott.freeserve.co.uk/casle/5.pdf (accessed 30 May 2007)

Powderham, A. (2002) 'The observational method – learning from projects', Proceedings of the Institution of Civil Engineers, *Geotechnical Engineering*, vol. 115, no. 1, January, accessed via http://www.tunnels.mottmac.com/search/?query=powderham (accessed 3 September 2007)

Roe, M. (2007) 'FIDIC and recent infrastructure developments', Pinsent Masons Press Article, available online at: www.pinsentmasons.com/media/1116563774.htm (accessed 30 May 2007)

Seppala, C.R. (2001) 'New standard forms of international construction contract', *International Business Lawyer*, February, pp. 60–65

Thomas, C., Hughes, S. and Glover, J. (2006) *Understanding the New Fidic Red Book: A Clause by Clause Commentary*, Sweet & Maxwell

Totterdill, B.W. (2006) *FIDIC Users' Guide: A Practical Guide to the 1999 Red and Yellow Books – Incorporating Changes and Additions to the 2005 MDB Harmonized Edition*, Thomas Telford

Useful websites

A network of expert users of the FIDIC Contracts: www.fidic-net.org

Articles of the FIDIC forms: www1.fidic.org/resources/contracts/#FIDIC%20Publications

European International Contractors: www.eicontractors.de

Institute of Civil Engineers
  www.ice.org.uk/knowledge/contracts_conditions_of_contract.asp

18 Case study: Heathrow Terminal 5

18.1 Introduction

The BAA Heathrow Terminal 5 (T5 henceforth) is one of Europe's largest and most complex construction projects. T5 was approved by the Secretary of State on 20 November 2001 after the longest public inquiry in British history (46 months) and when completed in March 2008 it will add 50% to the capacity of Heathrow and provide a spectacular gateway into London. Designed and engineered by Arup, with architects Richard Rogers Partnership and steel manufacturer Severfield Rowen, T5 has been described as engineering of Brunellian proportions.

The project includes not only a vast new terminal and satellite building but nine new tunnels, two river diversions and a spur road connecting to the M25; it is a multidisciplinary project embracing civil, mechanical, electrical systems, communications and technology contractors with a peak monthly spend of £80 million employing up to 8,000 workers on site. The construction of T5 consists of 18 main projects divided into 140 sub-projects and 1,500 'work packages' on a 260 hectare site.

Phase 1 construction of T5 is programmed for five years and can be broken down into five key stages:

1. *Site preparation and enabling works* (2 July–3 July) – preparing the site for major construction activity. The work included a significant amount of archaeological excavation, services diversions, levelling the site, removing sludge lagoons and constructing site roads, offices and logistics centres;
2. *Groundworks* (2 February–5 February) – includes the main earthworks, terminal basements, connecting substructures and drainage and rail tunnels;
3. *Major structures* (3 November–6 November) – the main terminal building (concourse A), first satellite (concourse B), multi-storey car park and ancillary structures;
4. *Fit out* (5 February–7 September) – significant items of fit-out include building services, the baggage system, a track transit people-mover system and specialist electronic systems;
5. *Implementation of operational readiness* (7 October–8 March) – ensuring phase 1 infrastructure is fully complete and that systems are tested, staff trained and procedures in readiness for operation in spring 2008.

Phase 2, which includes a second satellite and additional stands, will start after 2006 when the residual sewage sludge treatment site will be vacated. When completed in 2010, the two phases will enable Heathrow to handle an additional 30 million passengers per year (BAA T5 fact sheet, The Key Stages of Terminal 5).

18.2   Project management philosophy

The project management approach was developed based on the principles specified in the *Constructing the Team* (Latham, 1994) and *Rethinking Construction* (Egan, 1998) but went further than any other major project with two underlying principles:

1.   The client always bears the risk – no matter which procurement option is chosen;
2.   Partners are worth more than suppliers – BAA has developed an integrated project team approach.

The history of the UK construction industry on large-scale projects suggested that had BAA followed a traditional route T5 would end up opening two years late, cost 40% over budget with six fatalities (Riley, 2005). This would have been unacceptable to BAA as their funding is determined by five-yearly reviews of landing charges by its regulator who allows BAA a set rate of return, but in order to satisfy shareholders BAA are required to beat that. 'Massive cost overruns would have wrecked the company's reputation and sent the share price plummeting' (Riley, M. quoted in Wolmar, 2005).

Significantly BAA expected a high degree of design evolution throughout the project in order to embrace new technological solutions and changes in security, space requirements or facilities functionality. On such a complex project early freezing of the design solution was not realistic.

BAA realized that they had to rethink the client's role and therefore decided to take the total risk of all contracts on the project. Under traditional contracts (JCT and ICE forms) the parties are reactive and manage the effect (the consequences) resulting in claims where up to 40% of the total cost of the claims could be paid to quantity surveyors (QS) and lawyers. BAA thought differently and introduced a system under which they actively managed the cause (the activities) through the use of integrated teams.

This strategy was implemented through the use of the T5 Agreement under which the client takes on legal responsibility for the project's risk. In effect, BAA envisaged that all suppliers working on the project should operate as a virtual company. Executives were asked to lose their company allegiances and share their information and knowledge with colleagues in other professions.

BAA's aim was to create one team, comprising BAA personnel and different partner businesses, working to a common set of objectives by the following means:

●   The T5 Agreements with suppliers do not specify the work required, they are a commitment from the partners and statements of capability, capacity and scope to be provided;
●   The organization is based on the delivery of products, seen as operational facilities, not a set of buildings;
●   BAA has selected the best people to suit the project's needs including 160 highly experienced and capable professionals from other organizations;
●   By using collaborative software key information such as the timetable and the risk reports, the work scope is freely available to the integrated project team;
●   An 'organizational effectiveness director' with a team of 30 change managers provide training and support to implement the culture change required to work in an open and collaborative way (NAO, 2005).

Many of the suppliers involved in T5 were brought on-board at the earliest stages of the planning process. This enabled completely integrated expert teams to work together to identify potential problems and issues before designs were finalized and fabrication and construction began. As a result the teams of suppliers and consultants were in a position to add value while

---

Panel 18.1 BAA's approach to risk management

BAA's approach to risk management has been a key factor in keeping the project on budget and ahead of schedule. Terminal 5 is being constructed under the T5 Agreement which means BAA acts as the prime client and accepts most of the risk. With this burden removed from contractors and suppliers, it enables everyone working on T5 to

- focus on managing the cause of problems, not the effects if they happen;
- work on truly integrated teams in a successful, if uncertain environment;
- focus on proactively managing risk rather than devote energy to avoiding litigation.

Source: BAA T5 fact sheet, *Risk Management*.

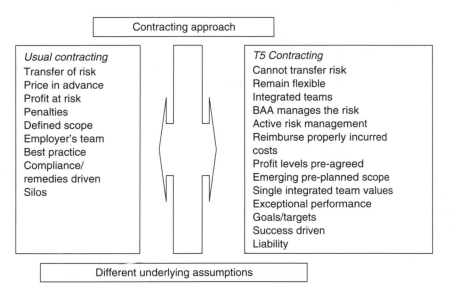

18.1 T5 Contracting philosophy (source: BAA document).

designing safe solutions within the time, quality cost and safety targets. This approach encouraged innovation; for example, the development of pavement concrete led to a 25% reduction in bulk materials required for the aircraft stands and pavement areas.

18.3   T5 Agreement

The T5 Agreement is a unique legal contract in the construction industry – in essence it is a cost reimbursable form of contract in which suppliers' profits are ring-fenced and the client retains the risk. It focuses in non-adversarial style on the causes of risk and on risk management through integrated team approaches. The reimbursable form of contract means that there are no claims for additional payments, and no payment disputes so far on the project (NAO, 2005). This move away from a lump-sum contract transfers a significant level of risk to the client and requires client-driven and client-owned systems to manage risk and facilitate collaborative behaviour amongst the project actors.

The T5 Agreement focuses on managing the cause and not the effect and ensures success in an uncertain environment. High performance levels and high benchmarking standards are demanded from all parties. Innovation and problem solving within the supply chain is actively encouraged. 'The idea is to have the best brains in all companies working out solutions to problems not working how best to defend their own corner' (T5's Commercial Director Matthew Riley quoted in Broughton, 2004).

For example, BAA benefited under the T5 Agreement as a result of the machinery and equipment suppliers pooling their purchasing power for cabling and other products at a discount resulting in savings up to 30% on some packages (Broughton, 2004).

BAA uses cost information from other projects, validated independently, to set cost targets. If the out-turn costs is lower than the target, the savings are shared with the relevant partners. This incentivizes the teams to work together and innovate. It is the only way to improve profitability: all other costs, including the profit margin, are on a transparent open-book basis (NAO, 2005). BAA takes precautions against risk of the target being too high through a detailed 'bottom-up' analysis by independent consultants.

The T5 Agreement creates a considerable incentive for performance. If the work is done on time, a third goes to the contractor, a third goes back to BAA and a third goes into the project-wide pot that will be paid only at the end (Douglas, 2005). Any payment is dependent on meeting milestones set in that agreement. Suppliers also benefit from ring-fenced profit and an incentive scheme that rewards both early problem solving and exceptional performance.

Panel 18.2 Designing value into the roof

As T5's main roof was a large element in the structure designing a cost-effective solution was critical to the project's success.

Richard Roger's Partnership's competition winning design envisaged a glorious expensive looking waveform roof supported on four rows of branched structural columns. This proved to be too complex and beyond the capability of the contractors – or in other words too expensive for BAA.

Critically in December 1999, a major value engineering exercise was undertaken involving all the key players: architects RRP, structural engineer Arup, steelwork contractor Severfield-Rowen, cladding specialist Schidlin and Hathaway roofing.

The development of the successful design became something of a saga with a solution developed through an iterative process. Buildability was a major issue due to the restrictions on site – at its highest point the roof towers 37 m above the apron; however, the airport's radar is in operation 2 m above that, thus prohibiting the use of cranes.

In the end the design team came up with a solution that satisfied all criteria: a single-span tied (or bowstring) arch supported high above the concourse on inclined structural columns. The roof is assembled on the ground in bays using 3,000 pre-assembled cassettes. The bays are then jacked up using the support abutments – in all, five lifts of three bays each and one single-bay lift.

Postscript: The positive approach to designing value into the roof is in complete contrast to approach used on the Scottish Parliament building where the value engineering exercises took place at a relatively late stage in the design and were considered ineffective.

Source: Pearson, *Building*, 2003.

Table 18.1 How the T5 documents fit together.

The document	What it is
T5 Agreement	The terms and conditions by which everyone working on T5 is bound
Supplemental agreement	The detail of the agreement which is signed by the suppliers. It defines the work they're doing on T5
Functional execution plan	The support required to enable projects to deliver
Sub-project execution plan	The team's plan of work
Work package execution plan	This is the breakdown of work by the supplier (combines preliminaries, specifications and drawings)
Supporting documents: Commercial policy Programme handbook Core processes and procedures Industrial Relations Policy	

Source: BAA document.

The final strand to the T5 Agreement is the insurance policy. BAA has paid a single premium for the multi-billion project for the benefit of all suppliers, providing one insurance plan for the main risk. The project-wide policy covered construction all-risk and professional indemnity.

Some key features of the T5 (Table 18.1) Agreement include the following:

- It is a legally binding contract between Heathrow Airport Ltd and its key suppliers.
- It says culture and behaviour are important. Innovatively, culture is specifically mentioned in the legal contract. The values – commitment, teamwork and trust – are key.
- It addresses risk and reward. BAA holds the overall delivery risk. Suppliers take their share of the financial consequences of any risk to the project and they also share in the financial rewards of success (like the project finishing on time and within budget).
- Risk payments, which would normally be costed into a supplier's quote, have instead gone into an incentive fund.
- Key project risks have been insured – loss or damage to property, injury or death of people, and, innovatively, professional indemnity for the project as a whole.

The T5 Agreement allows the project to adopt a more radical approach to the management of risk including early risk mitigation. Key messages include: 'working on T5 means everyone anticipating, managing and reducing the risks associated with what we're doing' (OGC, nd).

The legally binding contract centres on a 250-page handbook containing the same set of conditions for each supplier. Beneath that are a series of 2–3-page supporting documents defining particular capacities. These supporting documents can evolve as the working environment changes – flexibility is built in.

18.4 The approval process

BAA operates a five-stage approval process, which is based on the changing levels of risk during the development of the project. An important feature of this process is that BAA is prepared to move forward into the next stage without having completed production design. This dynamic streamlined decision process is in contrast to the linear Gateway process recommended by the UK Government for the procurement of public works (see Fig. 18.2). The NAO recommends that only clients who have strong in-house capacity as an intelligent client should use this form of procurement and management (NAO, 2005).

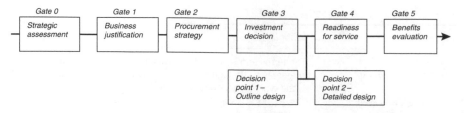

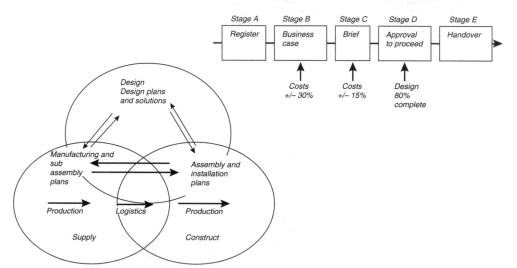

The OGC Gateway process offers a more 'linear' series of control points in a project's life

The T5 approach enables supply and production to proceed concurrently with design

18.2 OGC Gateway compared to BAA process (source: National Audit Office, 2005) – Copyright, National Audit Office, reproduced with kind permission.

18.5   Controlling the time, cost and quality

Keeping the project on schedule and within budget is obviously critical on a project this size. Traditionally the two elements tend to run separately, often in two separate sections – planning and costing.

T5 aims to be in the forefront of project control and is one of the first major users of the Artemis project management system in UK construction. The system is very robust and can show how each area of the project is performing relative to target, on both schedule and costs. A further key point of the Artemis system is that it can give information at programme or at individual project level or sub-project level. Cost and performance data can be analysed in various ways including the production of two highly useful indices, the 'schedule performance index' and the 'cost performance index' which are generated for all the levels and for each package (*New Civil Engineer*, 2004). This enables trends to be identified, highlights where performance is not as planned and most importantly, it enables informed management decisions to be made to keep the project on track.

The T5 project has a culture of right first time with no waste. To implement this theme, BAA has decided to pick up the costs if contractors get something wrong arguing that that they would be much more likely to own up quickly to the mistake and hence save a great deal more money (and time) when the mistake came to light anyway. However a contractor was not reimbursed for getting the same thing wrong twice; neither does it include fraud.

Overall quality governance is implemented via monthly review meetings for each of the 16 main projects and a monthly audit schedule workshop (Geoghegan, 2005).

On a practical level BAA organizes 'quality weeks' and 'quality benchmark' and 'interface centres' – small on-site showrooms in which supervisors can show the workers what standards are called for before the work starts.

18.6   Logistics

Despite being equivalent to London's Hyde Park in size, the T5 site itself is very physically constrained. To the North and South of the site are two of the world's most heavily utilized runways. Existing terminals are situated to the east and Europe's busiest motorway interchange (M25/M4) to the west. As a result, space on the site is at a premium; this together with the need to minimize construction traffic on local roads and the requirement for a single entrance and exit point through which all construction-related vehicles and people enter and leave presented the project team with immense logistical challenges.

Deliveries are made to the site only between 9.00 am and 5.00 pm through one entrance and exit, taking 170 deliveries per hour during that time during the five-year construction period (Black, 2005).

The project team's solution was based on methods used in factory-based manufacturing so that materials are brought to site only when the site is ready to receive them. This 'just in time' or 'pull' strategy is a first for a construction project of this scale. It is supported by extensive use of prefabrication and pre-assembled components and is facilitated by the use of two consolidation centres close to the main site.

This lean-thinking strategy has generated certain benefits including the following:

- eliminated the need for lay-down space for materials;
- increased reliability and efficiency in the use of materials which, in turn, has increased productivity levels (from a typical construction site average of 55–60% to an unprecedented 80–85%);

Panel 18.3 Proactive risk management – trial assembly roof abutment structure

The roof of the main T5 building comprises six sections each weighing 2,500 tonnes which had to be jacked up over a ten-month period.

To minimize any chance of mishaps and to ensure that the roof erection proceeded smoothly on site the T5 roof team, including designers, suppliers and fabricators pre-erected one of the 22 major roof abutment structures at the steel fabricator's (Severfield-Rowen) base near Thirsk in Yorkshire.

The pilot exercise proved to the design team that the erection method was workable and helped the construction team better understand the sequencing and tolerances required.

As a result the T5 team identified 140 significant lessons resulting in each having a risk-mitigation plan enabling faster construction on site.

This exercise cost BAA £4 million but saved three-months work on the Heathrow site, enabling delays that had previously arisen due to the wet winter of 2001/2002 to be recovered.

Source: NAO, 2005; BAA Terminal 5 DVD, 2004.

- driven out the traditional waste created by conventional practices;
- enabled BAA to uphold environmental commitments by reducing transport movements to and from the site while strictly controlling the timing of deliveries.

Project teams are required to plan their requirements for materials up to six weeks in advance. A software system called Project Flow has been developed which collates the team's demands and drives materials through the system. The materials are then delivered either just before or on the day they are required (BAA T5 Fact sheet Logistics).

18.7   The 3D project model

BAA set itself a target of using technology to reduce the total project cost of T5 by 10%. This was largely be achieved by creating a single 3D computer model that BAA and its project partners used to design, build and ultimately maintain the terminal building.

Major projects, involving multiple design teams, frequently suffer from poor collaboration and ambiguous design detail resulting in delays and increased costs. BAA used NavisWorks software to review 3D design data particularly where design models from different disciplines are brought together.

Conventionally, the architect designs the building and passes the CAD drawings over to the engineer. The engineer then draws the building over again for engineering analysis, as do the subcontractors, and the result is that the building and the elements within it are redrawn hundreds of times. BAA has proved in its research into the construction process that by the time the project gets to site these drawings are bound to contain inconsistencies, meaning that if different parts don't fit together they have to be reworked on site. The estimated cost of wasted time and materials alone is at least 10% of total project cost. If the costs of disruption to the programme are factored in, the figure is even higher.

The idea behind the single-project model is to derive an unambiguous set of data through the sharing of data. Using this approach the engineer never redraws the information; they reuse the architect's data and add to it. This approach drives out errors and improves efficiency.

A 3D model incorporating intelligent-object technology is being used at T5 to improve efficiency even more. This means that objects in the CAD drawing know what they are, and how they fit with other objects in the building.

The massive roof nodes connecting the roof structure are a good example of how the single-model environment worked in practice. Richard Rogers Partnership designed the node and passed it over to structural engineer Arup. The engineer used the architect's drawing to carry out structural analysis; RRP then modified the design to fit the analytical requirements using the same set of data. The model was passed to the steel fabricator Rowen Structures who used it to fine-tune the design of the parts of the node that had to be specially made. Finally they used the model to control the machinery that made the roof parts (Pearson, 2003).

Lessons learned at T5 are being disseminated to the rest of industry. The Construction Project Information Committee (CPIC) has published the *Code of Practice for Production Information*, which contains the processes and protocols used at T5 (CPIC, 2003).

18.8   The use of the NEC

Around 10% of the T5 value was procured under NEC contracts. The 70 first-tier suppliers, contractors and consultants were contracted under BAA's bespoke T5 Agreement. Under this arrangement each first-tier supplier was responsible for developing their supply chain to deliver the work. BAA recommended they use its version of the NEC *Engineering and Construction*

Panel 18.4 Construction facts

Capacity

Phase one opens in 2008 and includes the main terminal and the first of two satellite buildings. The second satellite is due for completion in 2011.
Passengers accommodated (total) 30 million per year

Dimensions

Area of T5 site	260 ha (around the same size as Hyde Park)
The main terminal	396 m long by 176 m wide by 40 m high (the interior space could accommodate around 50 football pitches)
The satellite building	442 m long by 52 m wide by 19.5 m high (bigger than Terminal 4)
	Twin river diversion 6 km of new river channels
Multi-storey car park	4,000 spaces

Bored tunnels

The T5 project includes a total of 13.5 km of bored tunnels

Construction stats

Length of site roads	6 km
No of tower cranes	30 at peak
Total volume of earthworks	6.5 million m³
Pavement quality concrete	335,000 m³ poured (4,700 m³ per week at peak)
Structural concrete	1.2 million m³ (14,000 m³ per week at peak)
Steel reinforcement	150,000 t
Structural steel	80,000 t (Wembley stadium has 23,000 t)
Number of lifts	175
Number of escalators	131
Number of site facilities	6 site compounds

Source: BAA T5 fact sheet, *Construction Facts*.

Contract for contracts with the thousands of second-tier suppliers – the only form recommended. This form was amended to work in line with the T5 Agreement and ensured that certain risks, that is, insurance excess deductibles were not passed down the supply chain.

BAA is also using various NEC contracts, particularly the *Professional Services Contract*, for around 50 direct relationships with consultants and other suppliers.

18.9 Role of the cost consultants

BAA selected a consultancy framework for cost consultancy on the T5 project comprising a collaborative joint venture of Turner & Townsend and EC Harris (the collaborative vehicle known as TechT). Both companies were selected under the same terms of commission and each provided 50% of the staff. On this project these two major consultancies became 'joined at the hip'. At its peak the cost consultancy team comprised 120 staff, approximately two-thirds of whom were QS(s). TechT provided both strategic and delivery services from inception through to the construction phase and has supported the BAA commercial team in

- designing the incentivized procurement strategy, business case and master plan;
- preparation of the project and resultant contract terms and conditions, processes and procedures and the implementation of a project-control system;
- ongoing cost and commercial management based on a target-cost philosophy within a cost-reimbursable incentivized framework.

Principal activities undertaken have included

- development of cost models allowing option appraisals within the master-planning phase and international benchmarking of airport using functionally based cost planning;
- development of indicators (operational and construction measures) to assist with target;
- identification of the performance standards within a value-management setting;
- development of the business case framework and instigating stakeholder-trading process;
- sensitivity analysis to test each options rates of return on investment.

The result of these activities supported the selection of an option and the procurement development of the functional brief within the budget parameters (www.turnerandtownsend.com [accessed 3 September 2007]). These activities give a real indication of the changing role of the cost consultant in the new environment. For an in-depth review of the role of the QS team on the Heathrow T5 project see chapter by author in Smyth and Pryke (2008).

18.10   Lessons learned

The key lessons learned from this significant project have been the following:

- The client always carries the risk – the key issue is the importance of managing risk and the proactive role taken by BAA in developing the T5 Agreement in which they take all the risk;
- Project management is a tool for risk and opportunity management not the other way round;
- Put risk management in the hands of those best able to manage the risk and adopt forms of contract that support a risk-management approach;
- Use of integrated supply teams with equality between all members has substantial benefits for all parties – this approach encouraged joint problem solving and innovation; the project was managed from open-plan offices on site, incorporating integrated management teams comprising BAA's staff with management teams from the key suppliers;
- Leadership and sponsorship at board level is vital;
- Using technology to cut 10% off the overall cost through the use of the single 3D project model;
- The considerable investment in temporary infrastructure without remaining value, for example, the £25 million temporary rail head to accommodate delivery of bulk raw materials;
- The materials consolidation centres in close proximity to the site enabled the implementation of a just-in-time strategy supporting and reducing the requirement for storage from 3 weeks to 3 days;
- The extensive use of prefabrication and pre-assembled components leading to an increase in productivity of between 10% and 15% when compared to the average building site;

Panel 18.5 Off-site prefabrication and pre-assembled components

Services – 60% assembled off-site in modules (5,000 modules based on 11 standard types); steel roof – 30 m sections; roof covering – based on 3 m x 6 m cassettes; traffic-control tower – the top 27 m of the tower including the visual control room constructed and partially fitted out 2 km from final location; steel reinforcement – 80% use of prefabricated cages; river diversions – 5 km constructed using precast concrete sections.

Advantages: increase in productivity; reduces overall programme time; improved quality of assembly; safer and quicker working in factory than on site; reduces risk of adverse weather; reduces on-site labour thus side-stepping skills shortages in South East of England.

Source: Pearson, *Building*, 2004.

- Aim to standardize and simplify, for example, aim to have eight types of light bulbs only;
- Specialist environmental team to monitor and control potential environmental impacts;
- Community liaison – aim to be a good neighbour with public exhibitions, regular contact with local villages and support for local facilities;
- Culture of benchmarking to measure performance based on ten industry key performance indicators (KPIs) developed in the Construction Best Practice Programme;
- Development of precise logistical strategy – this was an immense challenge;
- Accredited health-and-safety test centre (all new workers coming on to the site had to pass the construction skills certification scheme (CSCS) test within three months of starting on site);
- Development of a safety culture with a safety record four times better than the industry average – thus attracting skilled workers;
- Health awareness and promotion campaigns including free medicals;
- Time and money spent on planning is time and money well spent;
- Linking incentive payments to the supply teams to the achievement of visible milestones (70 milestones in total, for example diversion of the twin rivers and lifting the roof);
- Construction Training Centre producing 80 new modern apprentices per year;
- Seek excellent, highly experienced people to work on major projects and minimize confrontation through establishing trust and openness in working relationships.

18.11   Conclusion

The T5 project is the watershed in embracing the principles of lean construction in the UK and has required a complete change in the mindset and culture of the participants. The client has a huge role to play in the project success. Instead of writing into its contracts penalties for failure BAA accepted all the risk from the outset and guaranteed its suppliers an agreed margin thus sending out a positive message to the whole project team. This approach is at the opposite end of the spectrum when compared to the UK Government's preferred PFI/PPP model which puts as much risk as possible onto its contractors.

BAA is an informed client that knows how to run an airport and appreciates those who know how to build take the leadership role. It created a single entity harnessing the 'intellectual horse-power' working to get the job done rather than poring over contracts to find excuses. In return for its goodwill BAA demanded absolute transparency in the books of its suppliers in order to minimize waste (Douglas, 2005).

This approach created an environment in which all team members were equal. Furthermore, it encouraged problem solving and innovation to drive out all unnecessary costs, including claims and litigation, and drive up productivity levels.

BAA's enlightened approach created a collaborative environment which leads to the implementation of industry-best practices and world-class performance. This approach is particularly relevant to long-term projects with high risk and high complexity, valued at £200 million and above, but might not be so relevant for smaller, more straightforward projects.

So does T5 represent history in the making? In his 2005 lecture at the Royal Academy of Engineering, Andrew Wolstenholme, BAA's Project Director, confirmed that he believes it does and that the new approach to project management as set out in the T5 Agreement will help the industry change for the better. However it will require a massive culture change to become the norm.

Bibliography

Black, D. (2005) 'BAA's Lean Construction Innovations', LCI's 7th Annual Lean Construction Congress, 20–23 September, San Francisco, California, USA

Boultwood, J. (2005) 'Heathrow T5: A Case Study', NEC Annual Seminar, ICE London, 26 May

Broughton, T. (2004) 'T5 a Template for the Future: How Heathrow Terminal 5 Has rebuilt the building industry', Supplement within *Building* magazine, 27 May

Construction Project Information Committee (CPIC) (2003) *Code of Practice for Production Information*, CPIC

Douglas, T. (2005) 'Interview: Terminal 5 approaches take-off', *The Times*, Public Agenda Supplement, 6 September

Fullalove, S. (ed.) (2004) 'NEC helps BAA deliver Heathrow Terminal 5', *NEC Newsletter*, no. 30, August

Geoghegan, M. (2005) 'Quality Governance and Management of Heathrow T5 Construction', Meeting London Branch IQA, May

Lane, R. and Woodman, G. (2000) 'Wicked Problems, Righteous Solutions – Back to the Future on Large Complex Projects', International Group for Lean Construction Eighth Annual Conference (IGEC), Brighton, England, July

National Audit Office (2005) *Improving Public Services through Better Construction: Case Studies*, Report by the Comptroller and Auditor General, HC 364, Session 2004–2005, 15 March

New Civil Engineer (2004) 'NCE Terminal 5 supplement', February, *New Civil Engineer*

Office of Government Commerce (OGC) (no date) 'Managing Risks with Delivery Partners: A Guide for Those Working Together to Deliver Better Public Services', OGC

Pearson, A. (2003) 'T5 satisfying hells hounds wrestling with serpents', *Building* magazine, 25 July

Pearson, A. (2004) 'The big picture', *Building* magazine, 8 October

Riley, M. (2005) 'Interview', *Turner & Townsend News Issue 31*

Smyth, H. and Pryke, S. (2008) *Collaborative Relationships in Construction*, Blackwell Publishing

Wolmar, C. (2005) 'Project management at Heathrow Terminal 5', *Public Finance*, April 22

Woodman, G.R. *et al.* (2002) 'Development of a Design and Cost Optimisation Model for Heathrow Airport Terminal 5', Federal Aviation Administration Airport Technology Transfer Conference, Paper 71

Useful websites

www.airporttech.tc.faa.gov/att04/2002%t20TRACK%20P.pdf/P-71.pdf (Federal Aviation Administration Airport Technology Transfer Conference, 2002)

www.heathrowairport.com (BAA The T5 experience fact sheets)

www.london-sw-branch-iqa.org.uk/meetings.htm (Institute of Quality Assurance)

www.nao.org.uk/publications/nao_reports/04–05/0405364case_studies.pdf (National Audit Office Reports)

www.newengineeringcontract.com/newsletter/index.asp (New Engineering Contract Newsletters)

www.productioninformation.org/final/contents.html (Code of Practice for Production Information)

www.raeng.org.uk/news/publications/ingenia/issue22/Kimberley.pdf (Royal Academy of Engineering)

www.reformingprojectmanagement.com (Leading journal in lean construction)

www.turnerandtownsend.com (Case Studies: Terminal 5 Heathrow Airport – Role of the Cost Consultant)

Table of cases

Note: References such as '178–179' indicate (not necessarily continuous) discussion of a case across a range of pages, whilst those '265p17.2' refers to Panel 17.2 on page 265.

Index

Note: References such as '181–182' indicate (not necessarily continuous) discussion of a topic across a range of pages, whilst '116p8.2' denotes Panel 8.2 on page 116, '175t11.1' Table 11.1 on page 175 and '249f16.1' Figure 16.1 on page 249. Wherever possible in the case of topics with many references, these have either been divided into sub-topics or the most significant discussions of the topic are indicated by page numbers in bold. Because the entire volume is about quantity surveying and project management, and certain other terms (e.g. 'clients') occur constantly throughout the work, the use of these terms as entry points has been minimized. Information will be found under the corresponding detailed topics.